Asking Questions in Biology

Visit the *Asking Questions in Biology*, Fourth Edition Companion Website at **www.pearsoned.co.uk/barnard** to find valuable **student** learning material including:

- Instruction on using SPSS
- Instruction on using R
- Instruction on using Minitab 16
- Instruction on using Gensat 14
- AQB 4e Spreadsheet
- Sample Data Files

Fourth Edition

Asking Questions in Biology

A Guide to Hypothesis Testing, Experimental Design and Presentation in Practical Work and Research Projects

Chris Barnard, Francis Gilbert

School of Biology, University of Nottingham

and Peter McGregor

Cornwall College, Newquay

Benjamin Cummings
is an imprint of

Harlow, England • London • New York • Boston • San Francisco • Toronto
Sydney • Tokyo • Singapore • Hong Kong • Seoul • Taipei • New Delhi
Cape Town • Madrid • Mexico City • Amsterdam • Munich • Paris • Milan

Pearson Education Limited

Edinburgh Gate
Harlow
Essex CM20 2JE
England

and Associated Companies throughout the world

Visit us on the World Wide Web at:
www.pearsoned.co.uk

First published under the Longman imprint 1993
Second edition 2001
Third edition published 2007
Fourth edition published 2011

ISBN 978-0-273-73468-0
British Library Cataloguing-in-Publication Data
A catalogue record for this book is available from the British Library

Library of Congress Cataloging-in-Publication Data
Barnard, C. J. (Christopher J.)
   Asking questions in biology : a guide to hypothesis testing, experimental design and presentation in
practical work and research projects / Chris Barnard, Francis Gilbert, Peter McGregor. --4th ed.
       p. cm.
   Includes bibliographical references and index.
    ISBN 978-0-273-73468-0 (pbk. : alk. paper) 1.   Science--Methodology. 2.   Biology--Methodology.
I. Gilbert, Francis S. (Francis Sylvest), 1956-II. McGregor, Peter K. III. Title.
   Q175.B218 2011
   570.72--dc22
                                         2011008299
10   9   8   7   6   5   4   3
15   14   13   12   11

Typeset in 9.5/12.5pt Concorde BE by 35
Printed and bound by Ashford Colour Press, Gosport

Contents

4 Presenting information

Supporting resources

Visit **www.pearsoned.co.uk/barnard** to find valuable online resources

Companion Website for students containing:

- Instruction on using SPSS
- Instruction on using R
- Instruction on using Minitab 16
- Instruction on using Gensat 14
- AQB 4e Spreadsheet
- Sample Data Files

For more information please contact your local Pearson Education Sales representative or visit **www.pearsoned.co.uk/barnard**

Preface

Science is a process of asking questions, in most cases precise, quantitative questions that allow distinctions to be drawn between alternative explanations of events. Asking the right questions in the right way is a fundamental skill in scientific enquiry, yet in itself it receives surprisingly little explicit attention in scientific training. Students being trained in scientific subjects, for instance in sixth forms, colleges and universities, learn the factual science and some of the tools of enquiry such as laboratory techniques, mathematics, statistics and computing, but they are taught little about the process of question-asking itself.

The first edition of this book had its origins in a first-year undergraduate practical course that we, and others since, ran at the University of Nottingham for many years. The approach adopted there now also forms the basis for a more advanced second-year course. The aim of the courses is to introduce students in the biological sciences to the skills of observation and enquiry, but focusing on the process of enquiry – how to formulate hypotheses and predictions from raw information, how to design critical observations and experiments, and how to choose appropriate analyses – rather than on laboratory, field and analytical techniques per se. This focus is maintained in the fourth edition. However, as in previous editions, we have responded to a number of positive suggestions from people who have used the book, either as teachers or as students, which we think enhance further its usefulness in teaching practical biology generally.

Again, the largest change has been with respect to the presentation of statistical tests. In the third edition we replaced the previous hand calculations with boxes based on the procedures and output of the Statistical Package for the Social Sciences (SPSS) and a bespoke package we have written called AQB, based on Excel. However, the world moves on: in our view, the era of large, expensive statistical packages is over. They are increasingly unwieldy and unreliable as they try to cover more and more of the techniques being developed, bolting them onto the same creaking framework. Biologists who are serious about analysing their data now usually use the freeware package called 'R', obtainable from **www.r-project.org**. This is a collaborative project of biologists and statisticians worldwide, who write code to carry out the latest techniques.

Thus in the fourth edition we have decided to base the analyses on R, to encourage undergraduates to use it. Unlike most modern programs, R is not a 'point-and-click' process, but you have to write the commands in yourself. This means you need to know what you are doing, otherwise you will not get any meaningful results. This is a barrier to modern undergraduates who have become used to simpler ways of obtaining output from such programs. Using R will be much more useful to students in the longer term. Experience teaching it to undergraduates shows that they learn it almost as quickly as any other package, and so the 'barrier' is not as large as many think.

We have retained the SPSS boxes and also created parallel boxes in Minitab and Genstat, as well as the hand-calculation boxes from the previous editions, which can all be downloaded from the book website (**www.pearsoned.co.uk/ barnard**) if desired. The formulae and hand calculations remain because, as previously, we consider it important that the underlying arithmetic of the tests is understood.

The range of tests remains as in the third edition, i.e. it includes, among other things, repeated-measures designs, analysis of covariance, multiple regression and principal components analysis. This is a wider spectrum of designs than is likely to be encountered in a first course in experimental design, but caters for many kinds of data collected in field-course and final-year projects. The Test Finder, Quick Test Finder and Help sections should enable students to find what test they need to carry out, while simultaneously underlining the principles involved.

With increasing emphasis on the wider communication and public understanding of science, we have retained the sections on presenting information. We include giving talks or presenting posters at scientific meetings, and writing for broader non-specialist readers such as newspapers and magazines. We retain the sections on using on-line literature databases such as the World of Knowledge, and on the ethical implications of working with biological material, now an essential consideration in any study as legal regulation of biological experiments in both teaching and research becomes ever more stringent.

The book looks at the process of enquiry during its various stages, starting with unstructured observations and working through to the production of a complete written report. In each section, different skills are emphasised and a series of main examples runs through the book to illustrate their application at each stage.

The book begins with a look at scientific question-asking in general. How do we arrive at the right questions to ask? What do we have to know before we can ask sensible questions? How should questions be formulated to be answered most usefully? Chapter 1 addresses these points by looking at the development of test-able hypotheses and predictions and the sources from which they might arise.

Chapter 2 looks at how hypotheses and predictions can be derived from unstructured observational notes. Exploratory analysis is an important first step in deriving hypotheses from raw data, and the chapter introduces plots and summary statistics as useful ways of identifying interesting patterns on which to base hypotheses. The chapter concludes by pointing out that although hypotheses and their predictions are naturally specific to the investigation in hand, testable predictions in general fall into two distinct groups: those dealing with some kind of *difference* between groups of data, and those dealing with a *trend* in the quantitative relationship between groups of data.

The distinction between difference and trend predictions is developed further in Chapter 3, which discusses the use of confirmatory analyses. The concept of statistical significance is introduced as an arbitrary but generally agreed yardstick as to whether observed differences or trends are interesting, and a number of broadly applicable significance tests are explained. Throughout, however, the emphasis is on the use of such tests as tools of enquiry rather than on the statistical theory underlying them. Having introduced significance tests and some potential pitfalls in their use, the book uses the main worked examples to show how some of their predictions can be tested and hypotheses refined in the light of testing.

In Chapter 4, the book considers the presentation of information. Once hypotheses have been tested, how should the outcome be conveyed for greatest effect? The chapter discusses the use of tables, figures and other modes of presentation, and shows how a written report should be structured. The chapter then moves on to consider the presentation of material in spoken and poster paper formats, and how to recast written reports of results for a general, rather than specialist, readership.

At the end of the book are a number of appendices. These provide expanded self-test questions and answers sections based on the material in the previous chapters and some statistical tables for use in significance testing.

We said that the book had its inception in our introductory practical course. This course was developed in response to an increasingly voiced need on the part of students to be taught how to formulate hypotheses and predictions clearly and thus design properly discriminating experiments and observations. As we descend into ever more neurotically prescribed teaching and assessment procedures, the need for students to be given clear guidance on such aspects of their work becomes correspondingly greater. As always, both our practical teaching and the book have continued to benefit enormously from our ongoing and enjoyable interaction with our undergraduates. Their insights and enquiries continue to hone the way we teach, and have been the guiding force behind all the discussions in the book.

Finally, we should like to thank all the people who have commented on the book since its first appearance, and encouraged us to think about the further amendments we have made in this present edition. In particular, we thank Tom Reader for the generous amount of time he has spent discussing the book with us, commenting on drafts of many of the amendments and giving freely of his experience in and enthusiasm for the business of communicating science. Rufus Curnow at Pearson Education encouraged us to produce a fourth edition, and guided us as to how it should be modified. James Gilbert and Lucy Browning made useful changes to the R descriptions.

Francis Gilbert
Peter McGregor
May 2010

Acknowledgements

We are grateful to Blackwell Science Ltd. for permission to reproduce copyright material: Figs 2.7 and 3.3 modified from *Choosing and using statistics*, published by Blackwell Science Ltd (Dytham, C. 2003).

Companion website

Material is available on the Internet to support students and lecturers using this book, at the web address **www.pearsoned.co.uk/barnard**. This includes the package AQB mentioned in the book, and downloadable Microsoft Excel® sheets that enable the student or lecturer to carry out in Excel all of the statistical tests described and included within the book. The tests covered include χ^2, correlation and regression, and the specific and general hypotheses tested by analyses of variance, both parametric and non-parametric. The package forms a useful supporting resource to the text. There are also documents containing parallel boxes to those in the text, outlining how to carry out the tests and interpret the output from SPSS, Minitab and Genstat.

1 Doing science
Where do questions come from?

You're out for a walk one autumn afternoon when you notice a squirrel picking up acorns under some trees. Several things strike you about the squirrel's behaviour. For one thing it doesn't seem to pick up all the acorns it comes across; a sizeable proportion is ignored. Of those it does pick up, only some are eaten. Others are carried up into a tree, where the squirrel disappears from view for a few minutes before returning to the supply for more. Something else strikes you: the squirrel doesn't carry its acorns up the nearest tree but instead runs to one several metres away. You begin to wonder why the squirrel behaves in this way. Several possibilities occur to you. Although the acorns on the ground all look very similar to you, you speculate that some might contain more food than others, or perhaps they are easier to crack. By selecting these, the squirrel might obtain food more quickly than by taking indiscriminately any acorn it encountered. Similarly, the fact that it appears to carry acorns into a particular tree suggests this tree might provide a more secure site for storing them.

While all these might be purely casual reflections, they are revealing of the way we analyse and interpret events around us. The speculations about the squirrel's behaviour may seem clutched out of the air on a whim, but they are in fact structured around some clearly identifiable assumptions, for instance that achieving a high rate of food intake matters in some way to the squirrel and influences its preferences, and that using the most secure storage site is more important to it than using the most convenient site. If you wanted to pursue your curiosity further, these assumptions would be critical to the questions you asked and the investigations you undertook. If all this sounds very familiar to you as a science student, it should, because, whether you intended it or not, your speculations are essentially scientific. Science is simply formalised speculation backed up (or otherwise) by equally formalised observation and experimentation. In its broadest sense most of us 'do science' all the time.

1.1 Science as asking questions

Science is often regarded by those outside it as an open-ended quest for objective understanding of the universe and all that is in it. But this is so only in a rather trivial sense. The issue of objectivity is a thorny one and, happily, well beyond the scope of this book. Nevertheless, the very real constraints that limit human objectivity mean that use of the term must at least be hedged about with serious qualifications. The issue of open-endedness is really the one that concerns us here. Science is open-ended only in that its directions of enquiry are, in principle, limitless. Along each path of enquiry, however, science is far from open-ended. Each step on the way is, or should be, the result of refined question-asking, a narrowing down of questions and methods of answering them to provide the clearest possible distinction between alternative explanations for the phenomenon in hand. This is a skill, or series of skills really, that has to be acquired, and acquiring it is one of the chief objectives of any scientific training.

While few scientists would disagree with this, identifying the different skills and understanding how training techniques develop them are a lot less straightforward. With increasing pressure on science courses in universities and colleges to teach more material to more people and to draw on an expanding and increasingly sophisticated body of knowledge, it is more important than ever to understand how to marshal information and direct enquiry. This book is the result of our experiences in teaching investigative skills to university undergraduates in the life sciences. It deals with all aspects of scientific investigation, from thinking up ideas and making initial exploratory observations, through developing and testing hypotheses, to interpreting results and preparing written reports. It is not an introduction to data-handling techniques or statistics, although it includes a substantial element of both; it simply introduces these as tools to aid investigation. The theory and mechanics of statistical analysis can be dealt with more appropriately elsewhere.

The principles covered in the book are extraordinarily simple, yet, paradoxically, students find them very difficult to put into practice when taught in a piecemeal way across a number of different courses. The book has evolved out of our attempts to get over this problem by using open-ended, self-driven practical exercises in which the stages of enquiry develop logically through the desire of students to satisfy their own curiosity. However, the skills it emphasises are just as appropriate to more limited set-piece practicals. Perhaps a distinction – admittedly over-generalised – that could be made here, and which to some extent underpins our preference for a self-driven approach, is that with many set-piece practicals it is obvious *what* one is supposed to do but often not *why* one is supposed to do it. Almost the opposite is true of the self-driven approach; here it is clear why any investigation needs to be undertaken, but usually less clear what should be done to see it through successfully. In our experience, developing the 'what' in the context of a clear 'why' is considerably more instructive than attempting to reconstruct the 'why' from the 'what' or, worse, ignoring it altogether.

1.2 Basic considerations

Scientific enquiry is not just a matter of asking questions; it is a matter of asking the *right questions* in the *right way*. This is more demanding than it sounds. For a start, it requires that something is known about the system or material in which an investigator is interested. A study of mating behaviour in guppies, for instance, demands that you can at least tell males from females and recognise courtship and copulation. Similarly, it is difficult to make a constructive assessment of parasitic worm burdens in host organisms if you are ignorant of likely sites of infection and can't tell worm eggs from faecal material.

Of course, there are several ways in which such knowledge can be acquired: e.g. the Internet/World Wide Web, textbooks, specialist academic journals (many of which are now available electronically through licensed subscribers like universities and colleges), asking an expert, or simply finding out for yourself through observation and exploration.

These days, the first choice for browsing information is often the Internet/ World Wide Web. The advantages of such 'online' searching in terms of speed and convenience hardly need detailing here, but there *are* dangers, as we indicate later. A good way of accessing reliable scientific information like this is to use one or more of the professional Web-based literature databases, such as the Web of Knowledge (**wok.mimas.ac.uk**), PubMed (**www.ncbi.nlm.nih.gov/pmc/**) Google Scholar (**scholar.google.co.uk**) or BIOSIS (**science.thomsonreuters.com/ training/biosis/**). These search the peer-reviewed (and therefore quality-controlled) academic journals for articles containing information relevant to your request. Each of these provides tips on how best to use them, but a handful of basic ones is given in Box 1.1.

Whichever mode of acquiring information is preferred, however, a certain amount of background preparation is usually essential, even for the simplest investigations. In practical classes, some background is usually provided for you in the form of handouts or accompanying lectures, but the very variability of biological material means that generalised and often highly stylised summaries are poor substitutes for hard personal experience. Nevertheless, given the inevitable constraints of time, materials and available expertise, they are usually a necessary second best. There is also a second, more important, reason why there is really no substitute for personal experience: the information you require may not exist or, if it does exist, it may not be correct. The Internet/World Wide Web is a particular hazard here because of the vast amount of unregulated information it makes available, often dressed up to appear professional and authoritative. *Such material should always be treated with caution and verified before being trusted.* Where academic information is concerned, a first step might be to check the host site to see whether it is a recognised institution, like a university or an academic publisher; another might be to look for other research cited in the information, for instance in the form of journal citations (*see* section 4.3.1), which can be cross-checked. Entering the author's name into the search field of one of the web-based professional literature databases (Box 1.1) to see whether this person has a published research track record can be another approach.

Box 1.1	Searching online literature databases

Searchable online literature databases, like the Web of Knowledge, BIOSIS or Pub-Med, allow you to search for articles by particular authors, or on particular topics, or according to some other category, such as a journal title or research organisation. An example of the kind of search fields on offer, in this case for the Web of Knowledge, is shown in Fig. (i).

The key to using the search fields effectively lies in the precision with which you specify your terms: too general and you will be swamped with articles that are of little or no interest; too narrow and you will wind up with only one or two and miss many important ones.

Figure (i) A screen capture from the Web of Knowledge as it appeared in 2010. Like other similar sites, it is regularly updated, so the exact appearance of the search field screen changes.

To help with this, the search fields provide various means of linking terms so that searches can be focused (the AND, OR, NOT options – called 'operators' – in Fig. (i)). However, the process inevitably involves some compromises.

For example, suppose you were interested in steroid hormone secretion as a cause of immune depression in laboratory mice. You might start, seemingly reasonably, by typing '*steroid hormone* AND *immune depression* AND *laboratory mice*' into the 'Topic' search field in Fig. (i) and hitting the 'Search' button. Disappointingly, and rather to your surprise, this yields nothing at all – apparently nobody has published anything on steroid hormones and immune depression in mice. At the other extreme, a search for '*immune* AND *mice*' yields over 45,000 articles, a wholly unmanageable number, of which many can be seen at a glance to be irrelevant to your needs. Clearly, something between the two is what is required.

The reason the first search turned up nothing is not, of course, because nobody has published

anything on the topic, but because the search term was restricted to a very specific combination of phrases. It could well be that people have published on the effects of steroid hormones on immune depression in mice but didn't use the precise phrases selected. For instance, they may have reported 'depressed immune responsiveness' or 'depressed immunity', rather than 'immune depression', and referred to specific hormones, such as testosterone or cortisol, rather than the generic term 'steroid'. There are various ways of catering for this. In the Web of Knowledge, the form '*immun* SAME depress** AND *mice*' in the 'Topic' search field allows the system to search for any term beginning with 'immun' or 'depress', such as 'immune', 'immunity', 'immunocompetence', 'depression', 'depressed' and so on, thus picking up all the variants. The term 'SAME' ensures similar combinations of phrase are recognised, in this case, say, 'immune depression', 'depressed immunity' or 'depressed immune response'. Running the search again in this form yields around 450 articles, much better than zero or 45,000, but with quite a lot of them still redundant. If the search is specified a little more tightly as *'immun* SAME depress* AND mice AND hormone'*, however, it turns up around 40 articles, and all much more on target.

All the searchable databases use these kinds of approaches for refining searches, some very intuitive, some less so. One thing you will quickly notice, though, is that exactly the same search can turn up a different number and selection of articles depending on which database you are using – BIOSIS, for example, manages to find something under the initial over-specific search that drew a blank on the Web of Knowledge. For this reason, it is good practice to run searches on a selection of databases.

Using general-purpose search engines, like Yahoo! or Google, can often turn up information from the professional literature too, but just as often you're likely to get information from unregulated personal websites, or other sources of uncertain provenance. Taking received wisdom at face value can be a dangerous business – something even seasoned researchers can continue to discover, the famous geneticist and biostatistician R. A. Fisher among them.

In the early 1960s, Fisher and other leading authorities at the time were greatly impressed by an apparent relationship between duodenal ulcer and certain rhesus and MN blood groups. Much intellectual energy was expended trying to explain the relationship. A sceptic, however, mentioned the debate to one of his blood-group technicians. The technician, for years at the sharp end of blood-group analysis, resolved the issue on the spot. The relationship was an artefact of blood transfusion! Patients with ulcers had received transfusions because of haemorrhage. As a result, they had temporarily picked up rhesus and MN antigens from their donors. When patients who had not been given transfusions were tested, the relationship disappeared (Clarke, 1990).

Where at all feasible, therefore, testing assumptions yourself and making up your own mind about the facts available to you is a good idea. Indeed, science is often characterised as systematic scepticism – a demand for evidence for every assertion. It is impossible to draw up a definitive list of what it is an investigator needs to know as essential background; biology is too diverse a subject, and every investigation is to some extent unique in its factual requirements. Nevertheless, it is useful to indicate the kinds of information that are likely to be important. Some examples might be as follows:

Question   *Can the material of interest be studied usefully under laboratory conditions or will unavoidable constraints or manipulations so affect it that any conclusions will have only dubious relevance to its normal state or functions?*

For instance, can mating preferences in guppies usefully be studied in a small plastic aquarium, or will the inevitable restriction on movement and the impoverished environment compromise normal courtship activity?

Or, if nutrient transfer within a plant can be monitored only with the aid of a vital dye, will normal function be maintained in the dyed state or will the dye interfere subtly with the processes of interest?

| Question | *Is the material at the appropriate stage of life history or development for the desired investigation?* |

There would, for instance, be little point in carrying out vaginal smears on female mice to establish stages of the oestrous cycle if some females were less than 28 days of age. Such mice may well not have begun cycling.

Likewise, it would be fruitless to monitor the faeces of infected mice for the eggs of a nematode worm until a sufficient number of days have passed after infection for the worms to have matured.

| Question | *Will the act of recording from the material affect its performance?* |

For example, removing a spermatophore (package of sperm donated by the male) from a recently mated female cricket in order to assay its sperm content may adversely affect the female's response to males in the future.

Or, the introduction of an intracellular probe might disrupt the aspect of cell physiology it was intended to record.

| Question | *Has the material been prepared properly?* |

If the problem to be investigated involves a foraging task (e.g. learning to find cryptic prey), has the subject been trained to perform in the apparatus and has it been deprived of food for a short while to make it hungry?

Similarly, if a mouse of strain X is to be infected with a particular blood parasite so that the course of infection can be monitored, has the parasite been passaged in the strain long enough to ensure its establishment and survival in the experiment?

| Question | *Does the investigation make demands on the material that it is not capable of meeting?* |

Testing for the effects of acclimation on some measure of coping in a new environment might be compromised if conditions in the new environment are beyond those the organism's physiology or behaviour have evolved to meet.

Likewise, testing a compound from an animal's environment for carcinogenic properties in order to assess risk might not mean much if the compound is administered in concentrations or via routes that the animal could never experience naturally.

| Question | *Are assumptions about the material justified?* |

In an investigation of mating behaviour in dragonflies, we might consider using the length of time a male and female remain coupled as an index of the amount of sperm transferred by the male. Before accepting this, however, it would be wise to conduct some pilot studies to make sure it was valid; it might be, for instance, that some of the time spent coupled reflected mate-guarding rather than insemination.

By the same token, assumptions about the relationship between the staining characteristics of cells in histological sections and their physiological properties might need verifying before concluding anything about the distribution of physiological processes within an organ.

The list could go on for a long time, but these examples are basic questions of practicality. They are not very interesting in themselves but they, and others like them, need to be addressed before interesting questions can be asked. Failure to consider them will almost inevitably result in wasted time and materials.

Of course, even at this level, the investigator will usually have the questions ultimately to be addressed – the whole point of the investigation – in mind, and these will naturally influence initial considerations. Before we develop this further, however, there is one further, and increasingly prominent, issue we must address, and that is the *ethics* of working with biological material.

1.2.1 Ethical considerations

Because biological material is either living, or was once living, or is derived from something that is or was living, we are sensitive to the possibility that another living organism may be harmed in some way as a result of what we are doing. Of particular concern is the possibility that our activity might cause such an organism to suffer, physically or psychologically. We try very hard to avoid suffering ourselves because, by definition, it is extremely unpleasant, so the question arises as to whether we should risk inflicting it on another living being simply because we are interested in finding something out about it. This is not an easy question to answer, not least because of the difficulty of knowing whether species very different from ourselves, such as invertebrates, are capable of experiencing anything that might reasonably be called suffering in the first place. However, good science is mindful of the possibility, and works to various guidelines and codes of practice, some enforced by law, to give organisms the benefit of the doubt. While minimising the risk of suffering is important in itself, there is also a straightforward practical reason why we should take care of the organisms we use, whatever they may be, since any results we obtain from them could be affected if the organism is damaged or in some way below par.

Suffering may not be the only potential ethical concern. If material is coming from the field, for example, there could be conservation issues. Is the species concerned endangered? Is the habitat it occupies fragile? Are there unwelcome consequences for populations or habitats of removing material and/or returning it afterwards? Questions like this can lead to acute dilemmas. For instance, the fact that a species is becoming endangered may mean there is a desperate need for more information about it, but the very means of acquiring the information risks further harm.

As awareness of these issues increases, ethical considerations are beginning to play a more explicit role in the way biologists approach their work, not just in terms of taking greater care of the organisms they use, and being better informed about their needs, but at the level of how investigations are designed in the first place. Take sample size, for instance. Deciding on a suitable sample size is a basic problem in any quantitative study. It might involve an informal judgement on the basis of past experience or the outcome of other studies, or it might depend on power tests (*see* Box 3.14) to calculate a sample size statistically. Where there are ethical concerns, a power test would arguably be better than 'guesstimation' because it would provide an objective means of maximising the likelihood of a meaningful result while minimising the amount of material needed (a smaller sample would risk the outcome being swamped by random noise, while a larger one would use more material than necessary). But, of course, the ideal sample size indicated by the power test might demand more material than can be sustained by the source, or involve a very large number of animals in a traumatic experimental procedure. The value of proceeding then has to be judged against the likely cost from an ethical perspective, a task with considerable room for debate. Detailed discussion of these issues is beyond the scope of this book, but a good idea of what is involved can be found in Bateson (1986, 2005), who provides a digestible introduction to trading off scientific value and ethical concerns, and the extensive ethical guidelines for teachers and researchers in animal behaviour published by the Association for the Study of Animal Behaviour (ASAB) and its North American partner, the Animal Behavior Society (ABS) (see **asab. nottingham.ac.uk** or **www.animalbehaviorsociety.org** or each January issue of the academic journal *Animal Behaviour*). It is also well worth looking at the website of the UK National Centre for the 3Rs (**www.nc3rs.org**; the three Rs stand for the Replacement, Refinement and Reduction in the use of animals in research), a government-funded organisation dedicated to progressing ethical approaches to the use of animals in biology. For discussion of more philosophical issues, see, for example, Dawkins (1980, 1993) and Barnard & Hurst (1996). It is important to stress that, tricky as these kinds of decision can be, ethical considerations should *always* be part of the picture when you are working with biological material.

1.3 The skill of asking questions

1.3.1 Testing hypotheses

Charles Darwin once remarked that without a hypothesis a geologist might as well go into a gravel pit and count the stones. He meant, of course, that simply gathering facts for their own sake was likely to be a waste of time. A geologist is unlikely to profit much from knowing the number of stones in a gravel pit. This seems self-evident, but such *undirected* fact-gathering (not to be confused with the often essential descriptive phase of hypothesis development) is a common problem among students in practical and project work. There can't be many science teachers who have not been confronted by a puzzled student with the plea: 'I've collected all these data, now what do I do with them?' The answer, obviously, is that the investigator should know what is to be done with the data before they are collected. As Darwin well knew, what gives data collection direction is a working *hypothesis*. Theories and hypotheses are absolutely vital to science, otherwise 'we shall all be washed out to sea in an immense tide of unrelated information' (Watt, 1971). With them, 'the enormous ballast of factual information, so far from being about to sink us, is used to reveal patterns and processes so that we need no longer to record the fall of every apple' (Dixon, 2000).

The word 'hypothesis' sounds rather formal and, indeed, in some cases hypotheses may be set out in a tightly constructed, formal way. In more general usage, however, its meaning is a good deal looser. Verma & Beard (1981), for example, define it as simply:

> a tentative proposition which is subject to verification through subsequent investigation. . . . In many cases hypotheses are hunches that the researcher has about the existence of relationships between variables.

A hypothesis, then, can be little more than an intuitive feeling about how something works, or how changes in one factor will relate to changes in another, or about any aspect of the material of interest. However vague it may be, though, it is formative in the purpose and design of investigations because these set out to test it. If at the end of the day the results of the investigation are at odds with the hypothesis, the hypothesis may be rejected and a new one put in its place. As we shall see later, *hypotheses are never proven, merely rejected if data from tests so dictate, or retained for the time being for possible rejection after further tests.*

1.3.2 How is a hypothesis tested?

If a hypothesis is correct, certain things will follow. Thus if our hypothesis is that a particular visual display by a male chaffinch is sexual in motivation, we might expect the male to be more likely to perform the display when a female is present. Hypotheses thus generate *predictions*, the testing of which increases or decreases our faith in them. If our male chaffinch turned out to display mainly

when other males were around and almost never with females, we might want to think again about our sexual motivation hypothesis. However, we should be wrong to dismiss it solely on these grounds. It could be that such displays are important in defending a good-quality breeding territory that eventually will attract a female. The context of the display could thus still be sexual, but in a less direct sense than we had first considered. In this way, hypotheses can produce tiers of more and more refined predictions before they are rejected or tentatively accepted. Making such predictions is a skilled business because each must be phrased so that testing it allows the investigator to discriminate in favour of or against the hypothesis. While it is best to phrase predictions as just that (thus: *males will perform more of display* y *in the presence of females*), they sometimes take the form of questions (*do males perform more of display* y *when females are present?*). The danger with the question format, however, is that it can easily become too woolly and vague to provide a rigorous test of the hypothesis (e.g. *do males behave differently when females are present?*). Having to phrase a precise prediction helps counteract the temptation to drift into vagueness.

Hypotheses, too, can be so broad or imprecise that they are difficult to reject. In general the more specific, mutually exclusive hypotheses that can be formulated to account for an observation the better. In our chaffinch example, the first hypothesis was that the display was sexual. Another might be that it reflected aggressive defence of food. Yet another that it was an anti-predator display. These three hypotheses give rise to very different predictions about the behaviour and it is thus, in principle, easy to distinguish between them. As we have already seen, however, distinguishing between the 'sexual' and 'aggressive' hypotheses may need more careful consideration than we first expect. *Straw man hypotheses* are another common problem. Unless some effort has gone into understanding the material, there is a risk of setting up hypotheses that are completely inappropriate. Thus, suggesting that our displaying chaffinch was demonstrating its freedom from avian malaria would make little sense in an area where malaria was not endemic. We shall look at the development of hypotheses and their predictions in more detail later on.

1.4 Where do questions come from?

As we have already intimated, questions do not spring out of a vacuum. They are triggered by something. They may arise from a number of sources.

1.4.1 Curiosity

Questions arise naturally when thinking about any kind of problem. Simple curiosity about how something works or why one group of organisms differs in some way from another group can give rise to useful questions from which testable hypotheses and their predictions can be derived. There is nothing wrong with 'armchair theorising' and 'thought experiments' as long as, where possible, they are put to the test. Sitting in the bath and wondering about how migratory birds manage to navigate, for example, could suggest roles for various environmental cues like the sun, stars and topographical features. This in turn could lead to hypotheses about how they are used and predictions about the effects of removing or altering them. By the time the water was cold, some useful preliminary experiments might even have been devised.

1.4.2 Casual observation

Instead of dreaming in the bath, you might be watching a tank full of fish, or sifting through some histological preparations under a microscope. Various things might strike you. Some fish in the tank might seem very aggressive, especially towards others of their own species, but this aggressiveness might occur only around certain objects in the tank, perhaps an overturned flowerpot or a clump of weed. Similarly, certain cells in the histological preparations may show unexpected differences in staining or structure. Even though these aspects of fish behaviour and cell appearance were not the original reason for watching the fish or looking at the slides, they might suggest interesting hypotheses for testing later. A plausible hypothesis to account for the behaviour of the fish, for instance, is that the localised aggression reflects territorial defence. Two predictions might then be: (a) *on average, territory defenders will be bigger than intruders* (because bigger fish are more likely to win in disputes and thus obtain a territory in the first place) and (b) *removing defendable resources like upturned flowerpots will lead to a reduction in aggressive interactions*. Similarly, a hypothesis for differences in cell staining and structure is that they are due to differences in the age and development of the cells in question. A prediction might then be: *younger tissue will contain a greater proportion of* (what are conjectured to be the) *immature cell types*.

1.4.3 Exploratory observations

It may be that you already have a hypothesis in mind, say that a particular species of fish will be territorial when placed in an appropriate aquarium environment.

What is needed is to decide what an appropriate aquarium environment might be so that suitable predictions can be made to test the hypothesis. Obvious things to do would be to play around with the size and number of shelters, the position and quality of feeding sites, the number and sex ratio of fish introduced into the tank, and so on. While the effects of these and other factors on territorial aggressiveness among the fish might not have been guessed at beforehand, such manipulations are likely to suggest relationships with aggressiveness that can then be used to predict the outcome of further, *independent* investigations. Thus if exploratory results suggested aggressiveness among defending fish was greater when there were ten fish in the tank compared with when there were five, it would be reasonable to predict that aggressiveness would increase as the number of fish increased, *all other things being equal*. An experiment could then be designed in which shelters and feeding sites were kept constant but different numbers of fish, say 2, 4, 6, 8, 10 or 15, were placed in the tank. Measuring the amount of aggression by a defender with each number of fish would provide a test of the prediction.

1.4.4 Previous studies

One of the richest sources of questions is, of course, past and ongoing research. This might be encountered either as published literature (*see* Box 1.1) or 'live' as research talks at conferences or seminars. A careful reading of most published papers, articles or books will turn up ideas for further work, whether at the level of alternative hypotheses to explain the problem in hand or at the level of further or more discriminating predictions to test the current hypothesis. Indeed, this is the way most of the scientific literature develops. Some papers, often in the form of mathematical models or speculative reviews, are specifically intended to generate hypotheses and predictions and may make no attempt to test them themselves. At times, certain research areas can become overburdened with hypotheses and predictions, generating more than people are able or have the inclination to test. If this happens, it can have a paralysing effect on the development of research. It is thus important that hypotheses, predictions and tests proceed as nearly as is feasible hand in hand.

1.5 What this book is about

We've said a little about how science works and how the kind of question-asking on which it is based can arise. We now need to look at each part of the process in detail, because while each may seem straightforward in principle, some knotty problems can arise when science is put into practice. In what follows, we shall see how to:

- frame hypotheses and predictions from preliminary source material,
- design experiments and observations to test predictions,

- analyse the results of tests to see whether they are consistent with our original hypothesis, and

- present the results and conclusions of tests so that they are clear and informative.

The discussion deals with these aspects in order so that the book can be read straight through or dipped into for particular points. A summary at the end of each chapter highlights the important take-home messages, and the self-test questions at the end show what you should be able to tackle after reading the book.

Remember, the book is about asking and answering questions in biology – it is not a biology textbook or a statistics manual, and none of the points it makes are restricted to the examples that illustrate them. At every stage you should be asking yourself how what it says might apply in other biological contexts, especially if you have an interest in investigating them!

We suggest you get hold of a statistical package in order to follow the analyses. The two we use in detail in the book are AQB and R. AQB is a set of Excel sheets that can be downloaded free from **www.pearsoned.co.uk/barnard**. You can download the installation file of R free from **www.r-project.org**, and follow the simple instructions to install it on your computer. This book will enable you to run analyses in R; a simple and very clear guide to learning more about R is available in Zuur *et al.* (2009).

References

Barnard, C. J. & Hurst, J. L. (1996) Welfare by design: the natural selection of welfare criteria. *Animal Welfare* **5**, 405–433.

Bateson, P. (1986) When to experiment on animals. *New Scientist* **109**, 30–32.

Bateson, P. (2005) Ethics and behavioral biology. *Advances in the Study of Behavior* **35**, 211–233.

Clarke, C. (1990) Professor Sir Ronald Fisher FRS. *British Medical Journal* **301**, 1446–1448.

Dawkins, M. (1980) *Animal suffering: The science of animal welfare*. Chapman and Hall, London.

Dawkins, M. (1993) *Through our eyes only?: The search for animal consciousness*. Oxford University Press, Oxford.

Dixon, A. F. G. (2000) *Insect predator–prey dynamics: Ladybird beetles and biological control*. Cambridge University Press, Cambridge.

Watt, K. E. F. (1971) Dynamics of populations: a synthesis. In P. J. den Boer & G. R. Gradwell (eds) *Dynamics of populations*. PUDOC, Wageningen, pp. 568–580.

Verma, G. K. and Beard, R. M. (1981) *What is educational research? Perspectives on techniques of research*. Gower, Aldershot.

Zuur, A. F., Ieno, E. N. & Meesters, E. H. W. G. (2009) *A beginner's guide to R*. Springer, London.

2 Asking questions
The art of framing hypotheses and predictions

So far, we've discussed asking questions in a very general way. Simply being told that science proceeds like this, however, is not particularly helpful unless it is clear how the principles can be applied to the situation in hand. The idea of this chapter is thus to look at the development of the procedure in the context of various investigations that you might undertake in practical and project work. We shall assume for the moment that the material of interest is derived from your own observations. We shall start, therefore, with the problem of making observations and directing them in order to produce useful information.

2.1 Observation

2.1.1 Observational notes and measurements

When first confronted with an unfamiliar system, it is often difficult to discern anything of interest straight away. This seems to be true regardless of the complexity of the system. For instance, a common cause of early despair among students watching animals in a tank or arena for the first time is the mêlée of ceaselessly changing activities, many of which seem directionless and without an obvious goal. An equally common complaint is that the animals seem to be doing nothing at all worth mentioning. It is not unusual for *both* extremes to be generated by the same animals.

In both of the earlier mentioned cases, the problem almost always turns out to be not what the animals are or are not doing, but the ability of the observer to observe. This is because observation involves more than just staring passively at material on the assumption that if anything about it is interesting then it will also be obvious. On the contrary, good observation is a skill that needs practice to develop, and experience to hone to a high level. To be revealing, observations may need to be very systematic, perhaps involving manipulations of the material to see what happens. They may involve measurements of some kind since some things may be apparent quantitatively rather than qualitatively. In themselves, therefore, observations are likely to involve a certain amount of question-asking. Their ultimate purpose, however, is to provide the wherewithal to frame testable hypotheses and the discriminating predictions that will distinguish between the hypotheses.

To see how the process works, we shall first give some examples of observational notes and then look at the way these can be used to derive hypotheses and predictions. These examples, therefore, develop through the book from initial observational notes, through framing and testing hypotheses, to producing a finished written report. The examples are based on the kinds of preliminary notes made during practical, field course and research project exercises by our own students at various stages of their undergraduate training. They come from four different fields of study, but their common feature is that they provide scope for open-ended investigation and hypothesis testing. Of course, the fact that we happen to have selected these particular examples to illustrate the process is irrelevant to the aim of the book, as the range of other examples running through it amply demonstrates. What emerges from the examples applies with equal weight in all branches of biology, from molecular genetics and cell biology to psychology and comparative anatomy.

Example 1

Material: Samples of leaves collected in the field from early-successional (dandelion, *Taraxacum officinale*; plantain, *Plantago lanceolata*; poppy, *Papaver rhoeas*), mid-successional (clover, *Trifolium repens*; ox-eye daisy, *Leucanthemum vulgare*) and late-successional (dogrose, *Rosa canina*; ragwort, *Senecio jacobaea*; blackthorn, *Prunus spinosa*; goldenrod, *Solidago canadensis*) plant species; graph paper; binocular microscope.

Notes: Collected leaf samples show a lot of variation in damage. Some leaves have several semicircles eaten in from the edges; some have numerous holes through the tissue, many brown round the edges; others show extensive damage with most of the leaf missing in many cases and damage to the stems and twigs they are on. There is considerable size variation in the leaves. Look at the size range in undamaged leaves and divide into three size classes. Measure size by drawing round whole leaves on graph paper and counting the squares within the boundary. Count number of leaves with different kinds of damage in each size class: 'small' leaves – 41 with small hole damage, four with marginal damage, 17 with severe damage; 'medium-sized' leaves – 12 with small holes, 29 with marginal damage, 34 with severe damage; 'large' leaves – two with small holes, 22 with marginal damage, 40 with severe damage. One obvious possibility is that size reflects height off the ground and the effects of different kinds of herbivore, perhaps mainly slugs and snails on the smaller leaves, caterpillars and other insects on the medium-sized (shrubs, bushes?) leaves and maybe cattle or

deer on the largest. It also looks as if the smaller (low-growing?) leaves have much more damage than the larger ones (84 per cent showing some damage, compared with 47 and 23 per cent in the other two categories). Maybe this is just because bigger plants have more leaves, so more escape damage. While examining the samples, notice several other things. Some leaves have tough 'skins', often with hairs on the surface; these are mainly from the medium-sized category (per cent with hairs visible to naked eye: small, 0; medium, 21; large, 9). Some of the bigger leaves smell strongly or have sticky or latexy sap when squeezed, but the last is also the case with what look like dandelion leaves. The stems of some of the medium-sized and bigger ones have thorns or sticky hairs. There's also more colour variation in these two categories, some leaves being reddish rather than green. It looks like the tougher, more strongly smelling samples generally have less damage than the others.

| Example 2 |

Material: Stained blood smears, vials of preserved ectoparasites, gut nematodes and faecal samples from live-trapped bank voles (*Myodes glareolus*), microscope with eyepiece graticule, clean microscope slides and coverslips, pipettes.

Notes: Looking at blood smears under a microscope, notice range of red and white blood cells. Some red cells in some of the smears have small stained bodies in them. These turn out to be a stage in the life cycle of a protozoan parasite, *Babesia microti*, which infects voles. Some slides seem to have much higher densities of red cells than others. The number of fleas and ticks in each vial varies a lot. Several voles didn't seem to have any, while some had a large number. Divide samples by age and sex, and do some counts of infected red cells; scan a roughly standard-width field along the graticule scale for each vole and count number of infected cells (adult males: 21, 45, 3, 0, 64; adult females: 16, 1, 13, 0, 0; juvenile males: 0, 5, 34, 0, 0; juvenile females: 0, 0, 0, 16, 0). Count ticks recovered from same groups (adult males: 0, 8, 3, 0, 7; adult females: 0, 1, 0, 0, 2; juvenile males: 2, 2, 0, 5, 3; juvenile females: 0, 0, 0, 2, 0). Smear some faecal samples onto slides and inspect under microscope. Lots of fragments of plant material and detritus. Some samples have clear oval objects which, on asking, turn out to be nematode eggs. Sometimes there are very large numbers of eggs, sometimes none. Too difficult to count all the eggs in each sample, so designate a rank score from 0 (no eggs) to 5 (more than 100 eggs). Scores for adult males: 3, 5, 5, 0, 3; for adult females: 2, 5, 0, 2, 2; for juvenile males: 2, 2, 4, 2, 1; for juvenile females: 0, 1, 0, 2, 1. On looking at the tubes of preserved worms from the same animals, notice that those with high egg scores do not always have more worms, but some have a greater range of worm sizes. Measure small samples of worms against scale on graticule (ranges for five males: 12–25 units, 9–18 units, 13–28 units, 17–22 units, 11–14 units; ranges for five females: 14–23, 13–19, 12–20, 13–30, 15–29 units).

Material: Vials of water containing suspensions of soil-dwelling nematodes from three sites differing in heavy metal and organophosphate pollution, microscope with eyepiece graticule, clean microscope slides and coverslips, pipettes, diagnostic key to common species morphologies.

Notes: Pipette 2×0.2-ml droplets of each sample onto a clean slide and examine under microscope. Count adult worms of identifiable species on each slide. Sample 1 (heavy metal polluted site) six apparent species, call A–F for the moment: numbers – A 27, B 5, C 17, D 3, E 32, F 2; Sample 2 (unpolluted) – B 43, C 4, D 15, F 18, plus four different species G 20, H 31, I 4, J 12; Sample 3 (organophosphate polluted site) – A 5, C 48, E 19, H 11. Juvenile worms also present but not readily identifiable to adult species. Nevertheless, numbers in each sample are 23 (Sample 1), 31 (Sample 2), 0 (Sample 3). Can also detect adult females with eggs. Number of females with eggs of each species in samples: Sample 1 – A 6, B 0, C 4, D 0, E 5, F 0; Sample 2 – B 13, C 0, D 2, F 6, G 5, H 11, I 0, J 1; Sample 3 – A 0, C 12, E 7, H 5. Samples contain quite a lot of detritus, some of which can be identified as decomposing nematode material but not related to species. Take ten arbitrarily chosen fields per slide and see how many contain at least one piece of decomposing material: Sample 1 – 5, Sample 2 – 3, Sample 3 – 7. Repeat for a further two droplets per sample. Number of adults in second set: Sample 1 – A 15, C 21, D 9, F 4; Sample 2 – B 31, C 6, D 21, F 11, G 16, H 24, J 3; Sample 3 – A 8, C 35, D 3, E 14, H 6. Juveniles per sample – 17

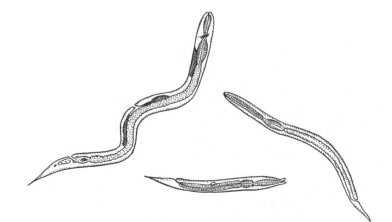

(Sample 1), 19 (Sample 2), 5 (Sample 3). Females with eggs: Sample 1 – A 3, C 4, D 0, F 0; Sample 2 – B 10, C 0, D 8, F 4, G 4, H 7, J 0; Sample 3 – A 0, C 11, D 0, E 5, H 0. Number of fields per slide with decomposing material: Sample 1 – 3, Sample 2 – 3, Sample 3 – 9.

Example 4

Material: Stock cage of virgin female and stock cage of virgin male field crickets (*Gryllus bimaculatus*), two or three 30×30-cm glass/Perspex arenas with silver sand substrate, dish of water-soaked cotton wool and rodent pellets, empty egg boxes, assorted colours of enamel paint, fine paintbrush, paint thinners, rule, bench lamps.

Notes: Females are distinguished from males by possession of long, thin ovipositor at the back. Put four males into an arena. After rushing about, males move more slowly around the arena. When they meet, various kinds of interaction occur. Interactions involve variety of behaviours: loud chirping, tapping each other with antennae, wrestling and biting. Interactions tend to start with chirping and antenna tapping, and only later progress to fighting. Count number of encounters that result in fighting (15 out of 21). Put in three more males so seven in total and count fights again (8 out of 25). Take out males and choose another five. Take various measurements from each male (length from jaws to tip of abdomen, width of thorax, weight) and mark each one with a small, different-coloured dot of paint. Introduce individually marked males into arena. Count number of fights initiated by each male per encounter (red, thorax width 6.5 mm – 4/10; blue, width 7.5 mm – 8/11; yellow, width 7.0 mm – 8/10; silver, 6.0 mm – 3/12; green, 6.5 mm – 3/9). Continue observations and count number of encounters won by each male (win decided if opponent backs off) (red, 3 wins/6 encounters; blue, 4/5; yellow, 7/7; silver, 1/4; green, 4/8).

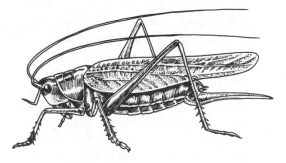

Introduce three sections of egg box to arena and leave males for 10 minutes. After 10 minutes, some males (blue, yellow, red) hiding under egg box shelters or sitting close to them. Males in or near boxes chirping frequently. Count number of encounters resulting in fight (12 out of 16). Count fights/encounter at different distances from burrows for each of the four males in turn – within 5 cm: 4 attempts in 6 encounters for yellow, 2/2 for red, 0/3 for silver, 4/5 for blue; between 5 and 10 cm: 5/11 (yellow), 2/3 (red), 1/5 (silver), 2/2 (blue);

10–15 cm: 3/7 (yellow), 4/8 (red), 0/2 (silver), 3/6 (blue); 15–20 cm: 2/9 (yellow), 1/6 (red), 1/5 (silver), 2/4 (blue); 20–25 cm: 1/8 (yellow), 1/12 (red), 0/2 (silver), 2/8 (blue). Introduce two females. Females move around the arena but end up mainly around the box sections. Show interest in males and occasionally mount. Other males sometimes interfere when female mounting particular male. Number of approaches to, and number of mounts with, different males: red, two approaches, no mounts; blue, two approaches, two mounts; yellow, three approaches, two mounts; silver, no approaches; green, one approach, one mount.

2.2 Exploratory analysis

Observational notes are, in most cases, an essential first step in attempting to investigate material for the first time. However, as the examples amply demonstrate, they are a tedious read and, as they stand, do not make it easy to formulate hypotheses and design more informative investigations. What we need is some way of distilling the useful information so that points of interest become more apparent. If we have some numbers to play with – and this underlines the usefulness of making a few measurements at the outset – we can perform some exploratory analyses.

Exploratory analysis may involve drawing some simple diagrams or plotting a few numbers on a scattergram, or it might involve calculating some *summary* or *descriptive statistics*. We shall look at both approaches shortly, using information from the various sets of observational notes. These sorts of analyses almost always repay the small effort demanded but, like much basic good practice in any field, they are often the first casualty of impatience or prejudgement of what is interesting or to be expected. It is always difficult to discern pattern simply by 'eyeballing' raw numbers, and the more numbers we have the more difficult it becomes. A simple visual representation like a scattergram or a bar chart, however, can turn the obscure into the obvious.

2.2.1 Visual exploratory analysis

There are several instances in the examples of observational notes where similar measurements were made from different kinds of material or under different conditions. For instance, infected blood cell counts were taken from adult and juvenile voles of both sexes, while fighting in male crickets was observed at different distances from artificial burrows. In both cases there seem to be some differences in the numbers recorded from different kinds of material (age and sex of host) or under different conditions (distance from a burrow). What do the differences suggest?

Eyeballing the blood cell data suggests some differences both between males and females and between adults and juveniles. A simple way to visualise this might be to total up the number of infected cells scored for each category of animal and present them in a bar chart (Fig. 2.1). If we do this, it looks as though

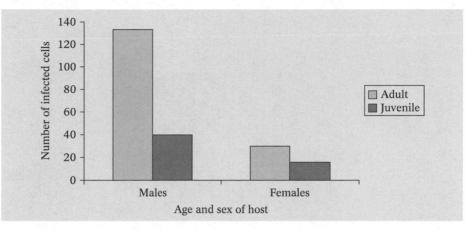

Figure 2.1 The total number of infected red blood cells in voles of different age and sex (*see* Example 2, Notes).

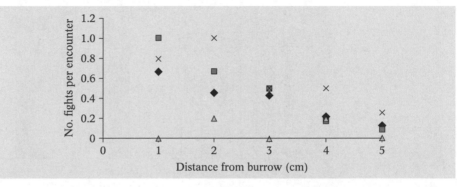

Figure 2.2 The number of fights per encounter between crickets at different distances from the nearest burrow; symbols represent individuals (*see* Example 4, Notes).

males generally have higher infection levels than females, regardless of age, and that adults have more infection than juveniles, regardless of sex. From this, we might be tempted to suggest that adult males are particularly prone to infection compared with other classes of individual. As we shall see later, however, we might want to be cautious in our speculation.

In the cricket example, we can see that the number of fights per encounter seems to vary with distance away from an artificial burrow. An obvious way to see how is to plot a scattergram of number of fights per encounter against distance. The result (Fig. 2.2) suggests that, while there is a fair spread of values at each distance, there is a tendency for more encounters to result in a fight when crickets are close to a burrow. It is important to bear in mind that, as each cricket was observed in turn, results for the different animals are independent of each other, and the trend is not a trivial outcome of a fight scored for one cricket also counting as a fight for his opponent. As we shall see later, this and other kinds of non-independence can be troublesome in drawing inferences from data.

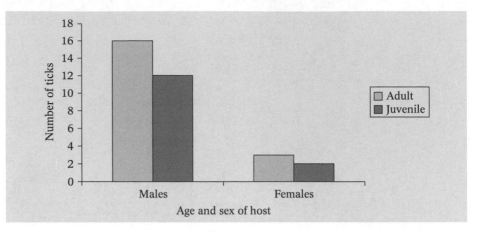

Figure 2.3 The number of ticks recovered from voles of different age and sex (*see* Example 2, Notes).

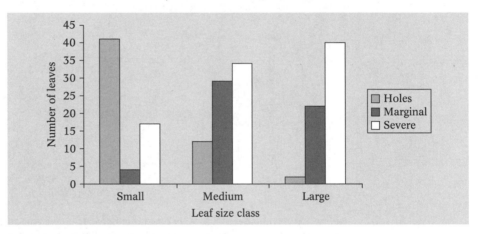

Figure 2.4 The number of small, medium and large leaves showing different kinds of damage by herbivores (*see* Example 1, Notes).

These are just two examples. We could do similar things with various other measurements in the observational notes. Figure 2.3, for instance, suggests that the number of *Babesia*-infected cells in different classes of host might have something to do with tick burden since both show the same broad association with class (*see* Fig. 2.1). Figure 2.4 hints at an association between the size of a leaf and the kind of damage it sustains. The Notes provide scope for other exploratory plots; try some for yourself.

Casting data in the form of figures like this is helpful not just because visual images are generally easier for most people to assimilate than raw numbers, but because they can expose subtleties in the data that are less apparent in numerical form. The plot of number of fights per encounter against distance from a burrow in Fig. 2.2, for instance, suggests that while the likelihood of fighting decreases further away from a burrow, there is considerable individual variation (the different symbols) within the trend. We shall see later that this variation leads to some interesting insights into the role of burrows in the behaviour of both male and female crickets.

The sorts of plots we have used so far are helpful in seeing at a glance whether something interesting might be going on. However, the data could be presented in a different way to make the figures more informative. It is clear from the scattergram in Fig. 2.2 and the raw data in the notes relating to Figs 2.1, 2.3 and 2.4 that there is a lot of variation in the numbers recorded in each case. Adult male voles, for example, did not all have high *Babesia* burdens; indeed one male didn't have any infected cells at all in the sample examined. This variability has at least two important consequences as far as exploratory plots are concerned. First it suggests that simply plotting totals in Figs 2.1 and 2.3 is likely to be misleading, because the totals are made up from a wide range of numbers. A large total could be due to a single large result, with all the rest actually being smaller than the results contributing to the other, lesser, totals, in which case our interest in the apparent differences between the bars in the figures might diminish somewhat. Second, variability in the data might obscure some potentially interesting tendencies in scattergrams. What we need, therefore, is a way of summarising data so that: (a) the interesting features are still made clear, but (b) the all-important variability is also presented, though in a way that clarifies rather than obscures patterns in the data. In short, we need some *summary statistics*.

2.2.2 Summary statistics

The usual way of summarising a set of data so as to achieve (a) and (b) above is to calculate a *mean* (average) or *median* value and then to provide as a measure of the variability the associated *standard error* (for a mean) or *confidence limits* (for a median).

Means and standard errors

The mean (often represented by $\bar{x}$ ('x-bar')) is simply the sum of all the individual values in the data set divided by the number of values (usually referred to as n, the sample size). Formally, the mean is expressed as:

$$\bar{x} = (1/n) \sum_{i=1}^{i=n} x_i$$

The expression $\sum_{i=1}^{i=n} x_i$ indicates that the first ($i = 1$) to the nth ($i = n$) data values (x) are summed (Σ is the summation sign) and can be expressed more simply as Σx. This summed value is then multiplied by $1/n$ (equivalent to dividing by the sample size n).

Since the mean is calculated from a number of values, we need to know how much confidence we can have in it. By 'confidence' we mean the reliability with which we could take any such set of values from the material and still end up with the same mean. A statistician would phrase this in terms of our sample mean ($\bar{x}$, the one we have calculated) reflecting the true mean (usually denoted μ) of the population from which the data values were taken. Suppose, for instance, we measured the body lengths of ten locusts caught in each of two different geographical areas and obtained the following results:

Length (cm)	
Area 1	Area 2
6.3	6.0
7.1	6.4
6.2	6.3
6.5	6.0
7.0	5.9
6.7	6.5
6.5	6.1
7.0	6.2
6.8	6.2
7.1	6.4
$\bar{x}$ 6.7	6.2

Then suppose that we had obtained the following instead:

Length (cm)	
Area 1	Area 2
8.3	8.0
5.1	5.4
7.2	5.3
5.5	6.5
8.0	8.4
5.7	5.5
5.5	7.1
8.0	5.2
8.8	5.2
5.1	5.4
$\bar{x}$ 6.7	6.2

In both cases, the mean body lengths from each area are the same, and we might want to infer that there is some difference in body size between areas. In the first case, the range of values from which each mean is derived is fairly narrow but different between areas. We might thus be reasonably happy with our inference. In the second case, however, the values vary widely within areas and there is considerable overlap between them. Now we might want to be more cautious about accepting the means as representative of the different areas. We can see that this is the case from the columns of numbers but we need some way of summarising it without having to present raw numbers all the time. We can do this in several ways. The most usual is to calculate the standard error, a quantitative estimate of the confidence that can be placed in the mean and which can be presented with it as a single number. The calculation is simple and is shown in Box 2.1.

Box 2.1 Standard deviations and standard errors

The *standard deviation* (usually abbreviated to s.d. or SD) measures the spread of actual data values around the mean, on the assumption that these data follow the *normal distribution* (*see* p. 58); it is a measure of the confidence you have that any particular data value will fall within a particular range (the mean + 1 s.d. and the mean − 1 s.d., hence the mean ± s.d.). The standard error of the mean, usually just called the *standard error* (abbreviated to s.e. or SE) measures the spread of multiple *sample means* around the true population mean. Normally you will only be taking a single sample, and hence the SE is an expression of the confidence you have that your sample mean falls within a particular range (mean ± s.e.) of the true population mean. Since sample means are almost always normally distributed, almost whatever the distribution of the raw data, it is always OK to cite a standard error with your sample mean. The calculations are as follows:

The standard deviation:

1. Calculate the sum of all the data values in your group (Σx).

2. Square the individual data values and sum them, giving (Σx^2).

3. Calculate $\Sigma x^2 - (\Sigma x)^2/n$ (remember that n is the sample size, the number of values in your set of data). This is actually a quick way of calculating the sum of squared deviations from the mean: $\Sigma(x - \bar{x})^2$. The deviations are squared so that positive and negative values do not simply cancel each other out.

4. Dividing by $n - 1$ gives the *variance* of the sample, an important intermediate quantity in many statistical tests.

5. Taking the square root of the variance gives the *standard deviation*.

Most scientific calculators will give you the mean of a set of numbers, and most will also give you the *standard deviation* (Box 2.1), usually represented as σ or s. If your calculator has both σ and σ_{n-1} buttons, it is the σ_{n-1} one that you want. The standard deviation will become important later, but for the moment we can simply use it to obtain the standard error. All we need to do is call up the standard deviation, square it, divide it by n and take the square root.

Whichever way you calculate the standard error (by hand or by calculator), it should be presented with the mean as follows:

$\bar{x} \pm$ s.e.

The ± sign indicates that the standard error extends to its value on either side of the mean. The bigger the standard error, therefore, the more chance there is that the true mean is actually greater or smaller than the mean we've calculated. We can see how this works with our locust data. Let's look at the two sets of values for Area 1, first calculating the x^2 value of the first example:

Locust	Body length (x)	x^2
1	6.3	39.69
2	7.1	50.41
3	6.2	38.44
4	6.5	42.25
5	7.0	49.00
6	6.7	44.89
7	6.5	42.25
8	7.0	49.00
9	6.8	46.24
10	7.1	50.41
$n = 10$	$\Sigma x = 67.2$	$\Sigma x^2 = 452.58$

The steps are then:

1. $\Sigma x = 67.2$

2. $\Sigma x^2 = 452.58$

3. $\Sigma x^2 - (\Sigma x)^2/n = 452.58 - (67.20)^2/10 = 452.58 - 451.58 = 1$

4. Divide by $n - 1 = 1/9 = 0.11$

5. Divide by $n = 0.11/10 = 0.01$

6. Take the square root $= \sqrt{0.01} = 0.11$

Thus the mean length of locusts in the first example for Area 1 is:

6.72 ± 0.11 cm

If we repeat the exercise for the second example, however:

1. $\Sigma x = 67.2$

2. $\Sigma x^2 = 471.18$

3. $\Sigma x^2 - (\Sigma x)^2/n = 471.18 - 451.58 = 19.6$

4. Divide by $n - 1 = 19.6/9 = 2.2$

5. Divide by $n = 2.2/10 = 0.22$

6. Square root $= \sqrt{0.22} = 0.47$

The mean is now expressed as:

6.72 ± 0.47 cm

We could leave the mean and standard error expressed numerically like this, or we could present them visually in a bar chart. If we opt for the bar chart, then the mean can be plotted as a bar and the standard error as a line through the top centre of the bar extending the appropriate distance (the value of the standard

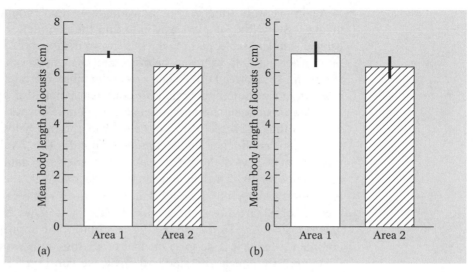

Figure 2.5 Mean (± s.e.) body lengths of locusts (using data from p. 24)

error) above and below the mean. Figure 2.5a, b shows such a plot for the two sets of example locust data.

Medians and confidence limits

An alternative summary statistic we could have used is the median. There are good statistical reasons, to do with the distribution of data values (*see* Chapter 3), why we may need to be cautious about using means and standard deviations (but not standard errors – see Box 2.1 – unless sample sizes are very low). The use of standard deviations in particular makes important assumptions about the distribution of the data from which they are calculated that may not hold in many cases. Using medians and confidence limits avoids these assumptions. Later, we shall see that statistical tests of significance can also avoid them. Finding the median is simple. All we do is look for the central value in our data. Thus, if our data comprised the following values:

   5   7   11   21   8   12   14

we first rank them in order of increasing size:

   5   7   8   11   12   14   21

and take the value that ends up in the middle, in this case 11. If we have an even number of values so there isn't a single central value, we take the halfway point between the two central values. Thus in the following:

   2   6   8   20   23   38   40   85

the median is 21.5 (halfway between 20 and 23). Note that the median may yield a value close to or very different from the mean. In the first sample, the mean is 11.1 and thus similar to the median. In the second sample, however, it is much greater at 27.58.

Again, we want some way of indicating how much confidence to place in the median. By far the simplest way is to find the confidence limits to the median using a standard table, part of which is shown in Appendix I. All we need to do is rank order our data values as before, count the number of values in the sample (n), then use n to read off a value r from the table. Normally, we would be interested in the r-value appropriate to confidence limits of approximately 95 per cent ('approximately' because, of course, the limits always have to be two of the values in the data set, one above the median and one below – if n is less than 6, 95 per cent confidence limits cannot be found). This r-value then dictates the number of values in from the two extremes of the data set that denote the confidence limits. Thus in our first sample data set, there are seven values. Reference to Appendix I shows that for $n = 7$, $r = 1$; the confidence limits to the median of 11 are therefore 5 and 21. However, if we had a sample of nine values (say 7, 11, 15, 22, 46, 67, 71, 82, 100) r for approximately 95 per cent confidence limits is 2, so for a median of 46, the limits would be 11 and 82.

As with the mean and standard error, we can represent medians and their associated confidence limits visually in a bar chart.

Another common way of expressing the distribution of values about the median is the interquartile range, which is the difference in values between the data point one-quarter of the way down the rank order of values in the sample and the point three-quarters of the way down. This is often presented in the form of a so-called 'box and whisker' plot (Fig. 2.6). This shows the median value as a bold bar within a box that represents the interquartile range. The 'whiskers' are

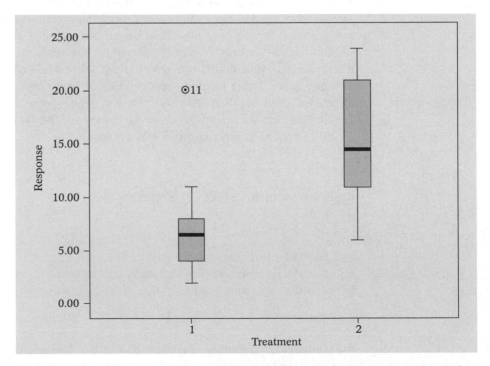

Figure 2.6 A box-and-whisker plot summarising the effect of a drug treatment and control on patient response. See text for details. Note the outlier indicated for the eleventh data point in Treatment 1.

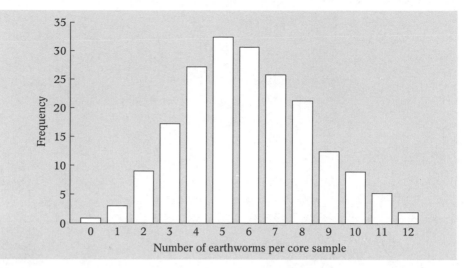

Figure 2.7 A bar chart frequency distribution of different numbers of earthworms recovered from core samples of soil (modified from Dytham, 2003).

not error bars in this case, but extend to the largest value in the sample that is within 1.5 interquartile ranges of the top or bottom of the box. Some computer statistical packages also present 'outlier' values as additional points on box and whisker plots; these can provide useful alerts to typographical errors in data entry.

Frequency distributions

Means and standard errors, medians and confidence limits, then, are two conventional ways of summarising central tendency and spread of values within data. As we have seen, they are particularly useful in making quick 'eyeball' comparisons between two or more data sets. Such 'eyeball' comparisons, however, are only one reason why visual summaries of data sets can be worth plotting. Another is to allow the features of a single set of data to be explored fully, perhaps to examine the distribution of values and decide on an appropriate method of confirmatory analysis (*see later*). The usual way of doing this is to plot a *frequency distribution* of the values (the number of times each occurs in the data set), either as a bar chart, like the means in Fig. 2.7, or as a histogram. Bar charts, in which the bars in the figure are separated by a small gap along the horizontal (*x*) axis, are used to plot the distribution of discrete values where there are sufficiently few of these to make such a plot feasible. Histograms, in which the bars are contiguous, are used to plot the distribution of classes of values (e.g. 1–10, 11–20, 21–30, etc.), usually where there is a wide and continuous spread of individual values. Figure 2.8 shows a range of frequency distribution histograms. It is obvious almost at a glance why we urged caution in using the mean as a general measure of central tendency, and why frequency distributions can be a crucial first step in deciding how to analyse and present data. In Fig. 2.8a the distribution is more or less symmetrical, with a peak close to the centre. Formally the peak is referred to as the *mode* and, in symmetrical, unimodal distributions like Fig. 2.8a, the mode and the arithmetic mean amount to the

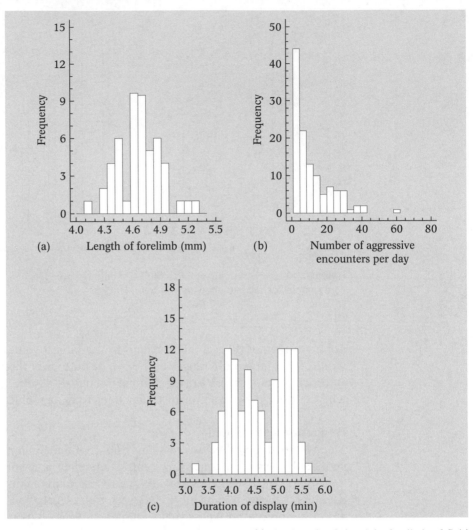

Figure 2.8 Histogram frequency distributions of: (a) the length of the right forelimb of field cricket (*Gryllus bimaculatus*) nymphs, (b) the number of aggressive encounters per day between male mice (*Mus musculus*) in an enclosure and (c) the duration of the aggressive displays of male Siamese fighting fish (*Betta splendens*) in pairwise encounters.

same thing. In these cases, therefore, the mean is an adequate measure of central tendency. In Fig. 2.8b, c, however, it is clear that the mean and mode are far from the same thing (in fact Fig. 2.8c has two modes and the distribution is said to be *bimodal*). Here, it makes little quantitative sense to use the arithmetic mean as a measure of central tendency. The shape of these distributions becomes extremely important when deciding on further analysis of the data, as we shall see shortly.

Of course, when plotting such distributions, one has first to decide on the number of categories into which to cast the data along the *x*-axis. Perhaps not surprisingly, there is no hard and fast rule about this. Dytham (2010), in his excellent book, advocates 12–20 categories as a rough rule of thumb, or, as an alternative, $\sqrt{n}$, where n is the number of data points in the sample. But, in the end, it's a matter of judgement and common sense.

Exploratory analysis, then, is a way of summarising data, visually or numerically, to make it easier to pick out interesting features. The calculations and plots we've suggested here can, of course, be produced by various computer packages rather than by hand. All the main statistical packages, like SPSS, Minitab, Statistica, GenStat, Statgraphics, GLIM, S-Plus and R, can do this, but many people prefer to enter their data into a general spreadsheet for organising and checking before transferring them into a statistical package. Microsoft's Excel is by far the most widely used of these spreadsheets, partly because it allows a range of exploratory analyses via simple programmable formulae, and easily produces data tables and plots (though see Box 2.3) that can be 'cut and pasted' into other documents or PowerPoint slides. Data in Excel can also be read directly into several different statistical packages, and so provide a convenient basis for further, more sophisticated analysis. We shall use Excel as our example spreadsheet package throughout the book; our example output uses the Microsoft Office 2003 and 2007 versions of the package. As our example statistical package, we have chosen to use R (*see* **www.r-project.org**), again because it is used widely in higher education and research and offers the range of tests required. The version we present in the screen captures is R 2.10.0. Box 2.3 summarises how to do the main kinds of exploratory analyses in Excel and R, but before looking at that, it is worth considering some general "do's and don'ts" in preparing data for analysis using computer packages. These are summarised in Box 2.2.

Box 2.2   Preparing data for analysis using computer packages

Although the modest data sets acquired during practical classes and projects can often be explored and analysed by hand without difficulty, it is now almost unheard of to do so. Instead, data are usually entered into one of the array of computer spreadsheet, database or statistical packages available commercially or via the Internet/World Wide Web, and analysed using these. Needless to say, these provide a quick and convenient means of dealing with data and are the virtually universal instruments of choice these days. However, as anyone who has ever used computer programs will know, it is important to maintain a healthy level of wariness when using them. The all-too-human potential for programmer and/or user error can lay serious traps for the unwary. On top of that, the immediacy of the results – click a button and there they are – does away with any need to understand what is actually going on. *When it comes to statistics,*

therefore, computer packages should be seen as a convenient adjunct to a proper formal training in the subject, not as a quick-fix substitute for it. It is thus impossible to over-emphasise the need to *check* and *double-check* that a package has done what you think it has when you carry out an operation – utterly meaningless rubbish can look perfectly plausible on superficial inspection.

While data can be entered into any program of your choice, we strongly recommend using Microsoft's spreadsheet package Excel, rather than a statistical package per se, for entering and organising your data ahead of analysis. Excel is a powerful program that allows data to be explored, prepared and analysed in many different ways. It comes as part of the standard installation of packages on many computers when they are purchased, so is widely used and supported. Data in Excel can also be imported easily into most statistical packages, so Excel is a convenient 'universal donor' when it comes to switching between different packages. The same is not true for many of the statistical packages themselves, where data often have to be converted into other formats (often Excel) before being exportable elsewhere, sometimes risking corruption on the way. It is a good idea, however, to save the finalised data set as a text file, with commas or tabs separating the values of the columns along each row, and with a header section explaining any codes in the data. This is valuable, because the formats used by commercial packages including Excel are always changing, but a simple text file will never be unreadable.

Excel's power and flexibility, however, is a two-edged sword. On the one hand it allows large data sets to be organised and explored easily; on the other, it provides a wealth of opportunities for error and confusion. Two of the commonest user problems are mistakes in using formulae to calculate new variables from existing ones (*see below*), and mistakes in using menu commands to reorganise data within a worksheet (most often inadvertently selecting only data visible on-screen, instead of the whole data set, when using the 'Sort' command, thus shifting some data out of line with the rest). The key to using Excel to prepare data for use in statistical packages is *'keep it simple'*.

Formatting data worksheets

When you boot up Excel, you are presented with the empty rows and columns screen in Fig. (i). This simple layout is all you need. Excel allows a vast range of row/column formats within its worksheets, and the tendency among students is to make liberal use of this capability, with header rows spanning several columns, worksheets subdivided into sections, textual notes and keys interspersed with data, data split across different worksheets and so on (e.g. Fig. (ii)). While all this is fine if the file is simply being used as a spreadsheet in its own right, it is completely unworkable if you want to import the data into a statistical package. As a general rule,

Figure (i) A newly opened Excel screen awaiting data entry.

Figure (ii) An example of an Excel worksheet format that would not be suitable for reading into a statistical package. Problems include: multiple data within cells, combinations of numerical data and textual comment within columns, and unsuitable column headings.

statistical packages expect data in simple column format, with each variable occupying a separate column and the first cell of the column containing the name of the variable. Each row then represents one study sample to which the data in each column cell along that row relate. Thus, suppose we had conducted an experiment looking at associations between anogenital sniffing (often a prelude to aggression) and circulating levels of the hormones testosterone (as a measure of sex hormone activity) and corticosterone (as a measure of stress levels) in male mice. Let's also suppose we had allocated the mice to four different experimental treatments, say different degrees of complexity of their group's home cage environment (e.g. (1) bare cage, (2) cage + nest material, (3) cage + nest material + nest boxes, (4) cage + nest material + nest boxes + shelves). The experiment consisted of introducing randomly chosen mice from eight different cages within each treatment into a clean empty cage and allowing them to interact one at a time with another randomly chosen mouse (different in each case) that had been kept on its own in an empty cage to standardise the social experience of the opponent across experimental subjects. Each pair of mice was allowed to interact for five minutes during which the number of anogenital sniffs initiated by the subject towards the opponent was recorded.

To set up our Excel data file for the experiment, we allocate one column to each of the variables in which we are interested. Thus our first column

might be *Mouse* and consists simply of numbers 1–32 so we know which mouse each row refers to. So we type 'Mouse' into the first cell in Column A and enter 1–32 in the successive cells below. Next, we want to note the experimental treatment for each mouse, so we head another column *Treatment*. Here, the best thing is to use a simple numeric code for each treatment, in this case '1' for bare cages, '2' for cage + nest material, and '3' and '4' for the other two cage types. Avoid using letters (e.g. A, B, C, D) or combinations of letters and numbers (e.g. T1, T2, T3, T4), because statistical packages sometimes treat these as text variables and won't allow them to be used in analyses. Luckily, R is not one of these. We can then complete three further columns for the number of sniffs initiated and the testosterone and corticosterone concentrations for each mouse, so the basic data set will appear as in Fig. (iii). Notice how the header row of variable names only occupies a single row, and each variable name is a *single* short word. If you *have* to use more than one word, use the dot (.) or an underscore (_) rather than a space in between the words; doing this avoids all sorts of potential problems when transferring the data to a statistical package, so we strongly recommend that you do.

We could, of course, add other measures if we wished, such as the body weight of each subject mouse, the time of day of each test, and so on, each in its own headed column. Depending on the nature of the variables, we could also calculate new ones within the file itself using Excel's formula capability (see also Boxes 2.3 and 2.4). Suppose, for instance, we wished to express anogenital sniffing in terms of the *rate* of sniffing rather than total number. Since all mice were observed for five minutes, this is simply a matter of dividing the number of sniffs by five to get the rate per minute. To do this, go to the first blank column after the existing data (here Column

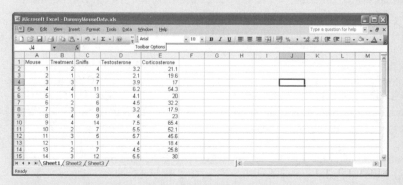

Figure (iii) Part of an Excel worksheet in a suitable format for reading into a statistical package.

F), head it *Sniffrate* (or similar) and click on the first empty cell below the heading (Cell F2). Now type an '=' sign, followed by C2/5 (so '=C2/5') and hit 'Return'. This produces the number 0.8 in the cell, which is the number of sniffs (from cell C2) divided by 5. There's no need to repeat this for every cell of the *Sniffrate* column; all we need to do is click on cell F2 again, then click on 'Edit' at the top of the screen, followed by 'Copy' (or just click on the 'Copy' symbol itself at the top of the screen). The cell is now highlighted by a shimmering dashed border, which indicates its contents have been copied to the clipboard. To complete the column, click on the cell below F2 and, holding down the 'Shift' key as you do it, the last cell in the column (here F25). The remainder of the column is now highlighted. Click 'Edit' and, this time, 'Paste' (or just click the 'Paste' symbol directly) to fill in the rates for each remaining cell. What Excel does here is to copy the actions of the formula you typed in (rather than the numerical content of Cell F2 itself) to each of the other cells you highlighted. An alternative method is to go to the bottom right-hand corner of the cell with the formula, and then either drag it down the column, or double-click the corner. The latter method is especially quick, but beware! Excel fills all cells in the column which has filled cells in the adjacent column. Thus if there are any missing values, the process will stop before it has filled the entire column. Always check that the column has been filled.

While calculating new variables like this in Excel is easy, we recommend that you keep it to a minimum. This is because worksheets with variables calculated using formulae may not be read properly by some statistical packages, or require special steps to ensure they are read properly. Since all these calculations can be done just as easily within statistical packages themselves, using menus rather than embedded formulae, our advice is to do them there rather than in Excel. If you have formulae in some Excel cells, when you have finalised the data you should highlight the entire data set (by clicking on the top left grey square), copy it and then right-click, choose 'Paste Special' and select 'Values' and 'Ok'. This will replace the formulae with the calculated values, which can be read properly by statistical packages.

Reading Excel data into R

To read the data in Fig. (iii) into R, you must save it into a text file first, because R will not read Excel files directly. As mentioned above, it is actually a good idea to keep your data in a simple text file so that you can never be caught out by changing data formats.

Make sure that all the columns have a variable name in the first row, with each name consisting of a single short word, or if more than one word, separated with an underscore or a dot. Make sure that all the values of a given variable are in a single column, and all the variables measured on a single sample/case/individual are on a single row. The variables can be true measurements, or codes for group membership (factors), either integer numbers or alphanumeric names. Use the powerful 'sort' functions in Excel to make sure that all the cells with an alphanumeric name for the same group contain identically spelled codes (e.g. 'green'), because R will treat variants as different groups (e.g. 'Green' and 'green'): R is case-sensitive in everything – you have been warned!

It is essential that somewhere (in the Excel file on a different sheet, or in a separate text file) you annotate or otherwise keep a detailed description of what your variable names mean, and what units they are in. This is the source of more grief than anything else – you will not remember what your variable names mean in six months' time!

Make sure that there are no empty cells in the data file; cells with missing values should all contain 'NA' (the signal to R that this value is missing).

With the data file open in Excel, choose 'File' and 'Save As...', and then select 'Text (Tab delimited) (*.txt)' from the drop-down list in the 'Save As Type' box. Give the file a simple single-word name, and click 'Save' 'OK' and 'Yes', noting where the file is being saved, i.e. to which folder.

CLOSE Excel! R may not be able to read the file while it is still open in Excel: when you exit Excel, reply 'No' to the question about saving changes.

Then start up R, which will open in a window and present you with a prompt '>'. We strongly recommend you keep sets of R commands in text files, which can be cut-and-pasted into R. Use a text editor such as NoteTab Light or Vim (**www.vim.org**), or R-specific text editors such as Tinn-R (see **www.sciviews.org/Tinn-R/**) or RWinEdt (an R package). All these are free to download from the Internet. The advantage of an R-specific text editor is that it will help you with a number of things about which R is very fussy, such as closing brackets correctly, or case sensitivity.

Type or paste the command that tells R to read in a data file ('read.table') chosen from a Windows Explorer window ('file.choose()'), into an object called a dataframe (which is a matrix of cases by variables). You can give it any name you like, but the shorter the better, since you may have to type it many times. The '<-' sequence means 'gets', i.e. the new dataframe is filled with values from the file. The 'header=T' bit makes sure the names of the variables are read from the first line of the file:

```
> dtfr1 <- read.table(file.choose(),
  header = T)
```

check it has been read in properly by typing/pasting:

```
> dtfr1
```

and all the data should be listed in columns. If there is an error, the most likely source is in the variable names – keep them simple!

```
> names(dtfr1)
```

will display all the variables R has read.

```
> str(dtfr1)
```

will tell you what R considers each column to be called, its nature (integer, number, factor), and the first few values of each column.

Then type/paste:

```
> attach(dtfr1)
```

This makes all the variable names available for use in commands. If you don't do this you will need to refer to the variables within a dataframe using the '$' operator:

```
> dtfr1$left
```

will list the values of the variable left in the dataframe dtfr1.

You can read in many data files and make many attach commands, but this confuses R if variables have the same name in different data sets – a common cause of errors for R users. The alternative is, whenever you issue a command, to start by telling R which dataframe you want it to use:

```
> with(dtfr1,{...})
```

where '...' is a command (see later); or to use the option in some (but not all) commands that specifies where the data are, e.g.:

```
> glm(tail ~ factor(colour),
  data = dtfr1)
```

You can always find out the syntax of any particular command by typing the command name preceded by '?':

```
> ?glm
```

and a window will open with all the details of this command; you will see straight away whether there is a 'data =' option.

If you make a mistake, then use the ^ (up button) on the number pad; this will scroll up through the commands, and you can modify one, press Return, and the new command will be enacted.

To calculate a new variable in R, you simply type in the formula e.g.:

```
> new.variable <- left - right
```

R assumes that you want the same number of values in your new variable (its *length*) as there are in the variables 'left' and 'right'. If 'left' and 'right' differ in length, then new.variable will be as long as the longest one, with the values of the shorter one recycled so as to match.

Logical operators are also very useful:

```
> new.variable <- x < 5
```

will result in a list of TRUE and FALSE values of the same length as x, evaluating each element of x as to whether it is less than 5.

R is hugely flexible in its ability to calculate, modify, summarise, select, code and recode variables. The commonest operators and functions that you might need are the following:

Operators

- `+ - * / ^` for plus, minus, multiply, divide, raise to the power
- `> >= < <= == !=` for greater than, greater or equal to, less than, less or equal to, equal to, not equal to
- `ifelse(test,action if true,action if false)`

Be careful with the difference between '<-', '=' and '==':

- `> x <- 5` sets a variable of length 1 (i.e. a scalar) to be 5
- `> x = 5` does the same thing
- `> x <- tail=5` generates an error
- `> x <- (tail=6)` sets `tail` and x to 6 and of length 1 (scalars)
- `> x <- (tail==6)` sets x to be a vector of the same length as `tail`, and containing TRUE or FALSE according to whether the equivalent values in `tail` = 6

Brackets have particular meanings in R:

- `> x[2]` the second value of the vector x
- `> x(2)` perform the function x with its argument = 2

- `> mean(x)` perform the function 'mean' on the vector x
- `> x[2,]` all the columns of the 2nd row of dataframe/matrix x
- `> x[,2]` all the rows of the 2nd column of dataframe/matrix x
- `> x[14:19,2:3]` rows 14–19 of columns 2–3 of dataframe/matrix x

Functions

There are many functions in R, but the most useful are:

```
log(x), log10(x), exp(x), sqrt(x)
```

One element of R that you will need to understand is creating vectors from numbers, using 'c' (for 'combine'):

```
> xlim=c(1,100)
```

creates a vector called `xlim` containing the numbers 1 and 100.

The rep function is also very useful:

```
> treatments <- rep(c(1:100),times=3)
```

gives a vector with the numbers 1 to 100 repeated three times.

Chapter 2 ('Essentials of the R language') of Crawley (2007) is a very useful summary of the flexibility and power of the R language.

Box 2.3   Exploratory analysis using computer packages

Various spreadsheet and database, and pretty well all statistical, packages written for computers will enable you to explore data sets and calculate standard summary statistics. The means of doing so are usually simplest, and the range of options greatest, in statistical packages, because calculating statistics is their stock-in-trade. In spreadsheets and databases they're often something of a frill, and can be tricky to get at.

Exploratory analysis using Excel

Excel allows various kinds of exploratory analysis, some very straightforward, others requiring more complicated procedures that highlight Excel's limitations on the statistical side. Again, our advice would be to use Excel as a simple spreadsheet for preparing raw data and reserve statistical treatment, even of an exploratory kind, for a proper statistical package. Some basic analyses that can be done in Excel, however, can be summarised as follows:

Calculating summary statistics in Excel

Most useful summary statistics can be calculated by using one or other of the menu functions at the top of the screen (see Box 2.2, Figs (i) and

Descriptive Statistics

Input

Input Range: $C$2:$F$41

Grouped By: ● Columns ○ Rows

☐ Labels in first row

Output options

○ Output Range:

● New Worksheet Ply:

○ New Workbook

☑ Summary statistics

☑ Confidence Level for Mean: 95 %

☐ Kth Largest: 1

☐ Kth Smallest: 1

OK

Cancel

Help

Figure (i) The 'Descriptive Statistics' dialogue box in Excel.

(iii)), or directly by typing in a simple formula. For example, using the sample data in Box 2.2, Fig. (iii), we could calculate a range of basic summary statistics for the columns or rows by clicking on 'Tools' (2003)/'Data' (2007) followed by 'Data Analysis'[†] and 'Descriptive Statistics'. This generates the dialogue box in Fig. (i) above. If we wish to obtain summary statistics for each of our columns of data in Box 2.2, Fig. (iii), we can tick the 'Summary statistics' and 'Grouped by: Columns' options in the box in Fig. (i), and enter the first and last data cells of the data we wish to include in the 'Input range:' box. For the data in Box 2.2, Fig. (iii), this would be cell C2 (the first data value in column C [the first cell is taken up with the name of the variable]), and cell F41 (we're not interested in summary statistics for the first two columns, because these are just coding variables). In Excel format, the range is entered as C2:F41. Clicking the 'OK' button now results in a list of summary statistics for each column of data. These include some statistics relating to the shape of the frequency distribution of values in the column, such as *kurtosis* and *skewness*, which we shall deal with in the next chapter.

[†] You may need to install the 'Analysis ToolPak' before 'Data Analysis' appears under the 'Tools'/'Data' menu. It is an Add-in – use the Help to enable you to install it.

Also included are all the 'standard' statistics commonly used to summarise data: the mean and standard error, the median and mode, the standard deviation and variance, and the range, sum and number of values. Selecting the 'Confidence' option in Fig. (i) will add the 95 per cent confidence limits to the mean to the list. Unless otherwise instructed, Excel will present the summary statistics in a new sheet of the active data workbook. Clicking on the 'Output Range' button and specifying a new cell in the current worksheet, however, will result in the summary being presented next to the data themselves.

Alternatively, summary statistics can be obtained individually by selecting an empty cell in the data sheet, clicking the 'Paste function' (f_x) button at the top of the screen and selecting from the 'Statistical' menu (note that Excel uses 'AVERAGE' for the mean). Irritatingly, this menu does not offer the standard error to the mean, but this can be calculated easily from the standard deviation using a simple formula (see Box 2.4). Indeed, such formulae are what the 'Paste' function itself uses; these can be seen by clicking on the cell containing the calculated summary statistic and looking in the '=' row above the data columns. Once people are familiar with them, the direct use of formulae usually becomes the preferred method, since it avoids having to click through the succession of options in the dialogue boxes. Box 2.4 shows some examples.

Using pivot tables to summarise by category

Often, we want to summarise according to categories of data within a data set, such as, for example, the different cage treatments in the mouse data in Box 2.2, Fig. (iii). This can be done very simply for total values using the so-called 'pivot table' facility. Click on 'Data' at the top of the screen (Box 2.2, Fig. (iii)) and select 'Pivot Table and Pivot Chart Report . . .'. Let's assume we're interested in knowing how the total number of sniffs varies with treatment in the data set in Box 2.2, Fig. (iii). Select 'Microsoft Excel list or database' as the source of data in the first pivot table dialogue box and click 'Next'. In the 'Range box' of the new dialogue box, enter the inclusive cells of the data set; here we enter B1:F41 including the variable names, note, because the pivot

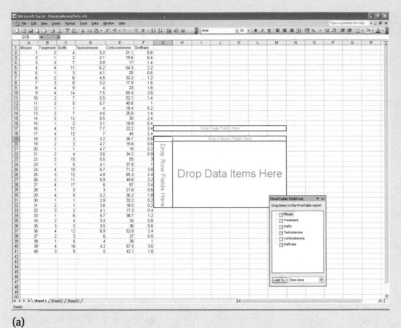

(a)

(b)

Figure (ii) (a) Setting up a 'pivot table' in Excel by dragging variable headings into the Rows and Columns box; (b) the resulting 'pivot table' in the worksheet.

table will use these to label the output. Clicking 'Next' produces the outline of the output table and a request to drag the 'field buttons' (corresponding to our selectable columns of data) on the right into the 'Row' or 'Data' compartments of the table (Fig. (ii)a). In this case, therefore, *Sniffs* is dragged into the 'Data' compartment (where it now says 'Sum of Sniffs'), since we're interested in variation in this with respect to treatment, and *Treatment* is dragged into the 'Row' compartment. Click 'Next' again, and select a target cell in the worksheet to receive the table, or select 'New worksheet' if you want it in a separate sheet, then click 'Finish' and the table will appear (Fig. (ii)b).

Sometimes, we might want to summarise by more than one category within our data (for instance, if we had recorded the number of sniffs by mice with respect to both their cage treatment *and* their social rank [dominant or subordinate] in their home cage). Where totals are concerned, this can be done easily once again using 'pivot tables'. Follow the procedure above, except this time specify three columns (the two categories and the variable of interest) in the 'Range box', and drag *both* category columns into the 'Row' compartment of Fig. (ii)a. Select a target cell for the table and click 'Finish', and a table of totals by one category within the other will appear.

Of course, it would be useful to be able to produce mean or median values along with their respective errors by category. Unfortunately, Excel provides no simple way of doing this, again underlining its limitations as a statistical package. While there are circuitous routes through which it can be done, it is really not worth the effort when the whole operation can be accomplished with a couple of commands in a statistical package like R.

Exploratory data plots in Excel

Using Excel to produce exploratory data plots like those in Figs 2.1–2.6 is straightforward, with the important exception of producing plots of mean/median values with their individual error bars. The figure-plotting menu does offer standard errors to means, but calculates a single standard error for the mean of the entire data set, not a separate error for the mean of each category within it, which is usually what is wanted. Again, there are long-winded ways around this, but it is far simpler to do the whole thing in a statistical package, which is what we recommend.

Plots of totals. Summary plots of totals by category, like Fig. 2.1, however, are easy to produce via the 'pivot table' function. The pivot table in Fig. (ii)b presents the kind of information we need. To produce a graph, all we need to do is click on the 'Chart wizard' option (the vertical coloured bars button) at the top of the screen, select 'Column', and the first chart subtype within that, and click 'Next'. Selecting the four treatment totals in the pivot table will enter their cell numbers in the 'Data range' box and produce a bar chart of them. Note that Excel defaults to scaling the x-axis as $1-n$ (here 1–4); if you wish to change this so the axis says, for example, '2, 4, 6, 8' or anything else, you need to enter your new labels into an empty column in your data set (so, here, just four cells of the column), right-click on the figure, select 'Source Data', then 'Series'. Put the cursor in the 'Category (X) axis labels' box and highlight the cells containing your desired labels; their specification will appear in the box. Clicking 'OK' will take you back to the now amended figure. Click 'Next' to edit as desired, or 'Finish' for the figure as it is. However you choose to edit the figure, we suggest you get rid of the default grid lines by right-clicking on the plotting area, clicking the 'Gridlines' option, and unchecking the relevant checkbox.

Scatterplots. To produce a scatterplot, like Fig. 2.2, highlight the two columns of data containing the X and Y data you want to plot, and click again on the 'Chart wizard' button and select 'XY (Scatter)'. Clicking on 'Finish' will then produce the scatterplot you want. If the X and the Y data are in columns that are not adjacent, then highlight the data in the first column, and press and keep down the Ctrl key while you highlight the matching data in the other column. Then press the 'Chart wizard' button and follow the procedure above.

Frequency distributions. Unfortunately, Excel does not provide a ready means of plotting frequency distributions. You will need to use a statistical package to do this (*see below*).

Exploratory analysis using R

R is a fully comprehensive statistical package, as opposed to a spreadsheet with some statistical facilities added on, like Excel. Consequently life is usually more straightforward when it comes to performing analyses, at least when they are fairly simple. It is far easier to import data into R from Excel (see Box 2.2) than to enter the data directly in R.

Calculating summary statistics in R

Nothing could be easier. To obtain summary statistics for any of the variables in your data set, simply issue the relevant command. The following functions return a single value (a scalar):

```
max(y), min(y), mean(y), median(y),
var(y)
```

or for dataframes you could type/paste:

```
colMeans(dtfr), colSums(dtfr),
rowMeans(dtfr), rowSums(dtfr)
```

but the easiest way to get these plus the interquartile range is:

```
> summary(y)
```

If 'y' is a dataframe, then we get the summary of all the variables in the dataframe. Variables with alphanumeric codes (factors) are counted.

If we want the same information for all the variables in the dataframe split into the different factor levels (codes) of a factor in the dataframe, then we just type/paste:

```
> tapply(dtfr, dtfr$factorname, mean)
```

The word 'mean' can be any function. We can use individual variables in `tapply` as well as dataframes.

The non-parametric equivalents are the minimum, the lower-hinge, the median, the upper-hinge and the maximum. These are obtained from:

```
> fivenum(y)
```

There are no built-in functions in R for standard deviation or standard error (for the difference, see Box 2.1), so they must be calculated:

```
> sqrt(var(x))
```

is the standard deviation

For standard error, we could write a function for use on many variables. This takes a generic form using 'x' to represent any variable given to the function as an argument:

```
> se <- function(x) sqrt(var(x)/length(x))
```

and then invoked by writing: `> se(y)`

To find out the sample size for different levels of a factor (or different integers of an integer variable such as 'year', use the 'table' command:

```
> table(Treatment)
> table(Treatment, year)
```

Exploratory data plots in R

Producing exploratory plots of data in R is fairly straightforward. However, we will want to use a text editor so that we save what we write, because we would not want to compose the commands from scratch every time.

Plots of totals. To plot total sniffs by treatment as we did above, for example, we need to save the totals in a vector and then plot it:

```
> sums <- tapply(Sniffs, Treatment, sum)
> barplot(sums, xlab="Treatment", ylab="Total sniffs")
```

Plots of means or medians. There is no built-in function in R for plotting means ± SE, and so we will have to create one (taken from Crawley, 2007: 462). Once it works, save it! We certainly don't want to have to redo this from scratch:

```
> plot.error.bars <-function(ymeans, yse, nn) {
  xv<-barplot(ymeans,ylim=c(0,(max(ymeans)+max(yse))),names=nn,
  ylab=deparse(substitute(ymeans)))
  g=(max(xv)-min(xv))/50
  for (i in 1:length(xv)) {
  lines(c(xv[i],xv[i]),c(ymeans[i]+yse[i],ymeans[i]-yse[i]))
  lines(c(xv[i]-g,xv[i]+g),c(ymeans[i]+yse[i],ymeans[i]+yse[i]))
  lines(c(xv[i]-g,xv[i]+g),c(ymeans[i]-yse[i],ymeans[i]-yse[i]))
  }}
>
```

The first line sets up the function itself, which takes the means, standard errors and labels as its arguments. Note that until we close all the brackets, continuation lines are indicated by the '+' rather than the prompt '>'. The second line sets up the object xv, the plot itself, which is careful to allow room for the error bars in its scaling of the *y*-axis: the phrase '`deparse(substitute(ymeans))`' turns the name of the variable into a label for the *y*-axis.

Then we just save the mean values, standard errors (using the routine developed above) and labels in appropriate variables, and plot using our new function:

```
> msniffs<-tapply(Sniffs, Treatment,
  mean)
> sesniffs<-tapply(Sniffs, Treatment, se)
> labels <- as.character(levels
  (Treatment))
> plot.error.bars(msniffs,sesniffs,
  labels)
```

We can plot the medians as boxplots, accomplished via declaring the values in Treatment to be factor levels first, and then issuing the plot command:

```
> Treatment <- factor(Treatment)
> plot(Treatment, Sniffs,
  xlab="Treatment",ylab="Total
  sniffs")
```

Scatterplots

To generate a scatterplot of the relationship between two variables, we type:

```
> plot(Testosterone, Sniffs,
  xlab="Testosterone", ylab="Sniffs")
```

We can control the axes scales using e.g. `xlim=c(lower,upper)`, so if we wanted to extend the x-axis from 0 to 8, and the y-axis from 0 to 25, then we would type:

```
> plot(Testosterone, Sniffs,
  xlab="Testosterone", ylab="Sniffs",
  xlim=c(0,8), ylim=c(0,25))
```

Symbols and colours can be manipulated at will. Suppose we wanted to use the default open circles for `treatment=1`, and closed circles for all other treatments, but colour-coding the closed symbols for each treatment. The various symbols and colours and their numerical codes are listed in Appendix IV. You do this by creating a numerical variable for the plot symbols (open circle = 1, closed circle = 16). We already have a variable with different integers for the treatments – `Treatment` itself – and we can use this for indicating the colours. We can also state the colours in words: 'col="red"', for instance. We manipulate the sizes of symbols using the 'cex =' option; a value of 1.2 would increase symbol size by 20%:

```
> plotchr <- ifelse(Treatment==1,1,16)
> plot(Testosterone, Sniffs, pch =
  plotchr, col = Treatment)
```

Frequency distributions.
To produce plots of the frequency distributions for variables, if the data are integers, we could first count their frequencies using the `table(int_var)` function; hence we type:

```
> freqs <- table(Sniffs)
> barplot(freqs,ylab="frequency",
  xlab="sniffs",col="red")
```

and this produces a red-coloured bar chart of the frequencies.

Alternatively, for continuous or integer variables, we could use the command:

```
> hist(sniffs)
```

If we wanted to suppress the irritating main heading, and control the break-points of the x-axis, then we could use:

```
> hist(Testosterone, main="",
  breaks=seq(-0.5,59.5,5))
```

Box 2.4   Some basic formulae for summary statistics in Excel

The following are some of the formulae commonly used to calculate summary statistics in Excel. The cell numbers in parentheses are, of course, arbitrary for the purposes of illustration:

Number of items in sample	'=COUNT(A2:A51)'
Arithmetic mean	'=AVERAGE(A2:A51)'
Mode	'=MODE(A2:A51)'
Median	'=MEDIAN(A2:A51)'
Variance	'=VAR(A2:A51)'
Standard deviation	'=STDEV(A2:A51)'
Standard error	'=(STDEV(A2:A51))/(SQRT(COUNT(A2:A51)))'
Interquartile range	'=QUARTILE(A2:A51,1)' for the first quartile (the value 25% up the data set when rank ordered),
and	'=QUARTILE(A2:A51,3)' for the third quartile (the value 75% up the data set)

As we emphasise in Box 2.2, using statistical packages on a computer will teach you nothing about the underlying assumptions and mechanics of particular tests. As far as the user is concerned, the package simply offers a magic box of tricks that throws up results when data are put into it. While we use our selected packages as the main presentational vehicle for statistical tests in the text – because, for better or worse, packages like these are what students will be recommended to use – we also include worked examples of the main kinds of test in the Appendices, so the reader can have some idea of what they actually do.

Of course, while interesting to do in their own right, exploratory analyses are useful only to the extent that they promote further investigation to confirm that what looks interesting at an exploratory level is still interesting when data are collected more rigorously and subjected to more thorough analysis. This brings us back to hypotheses and predictions, and leads to a consideration of *confirmatory statistics*.

2.3   Forming hypotheses

2.3.1 Turning exploratory analyses into hypotheses and predictions

Exploratory analyses are generally open-ended in that they are not guided by preconceived ideas about what might be going on. However, they are the first important step on the way to formulating hypotheses that do then guide investigation. As we have seen already, hypotheses can be very general or they can be specific. Both kinds can be generated from our observational notes.

Example 1

Plants and herbivores

The observations of leaf damage in the samples of plant species suggest that several factors are influencing the type and amount of damage. A possibility emerging from the Notes is that damage decreases with height off the ground and thus vulnerability to slugs. This can be framed as a readily testable hypothesis:

Hypothesis 1A *The type and extent of leaf damage reflects availability to slugs.*

from which some predictions for testing might be:

Prediction 1A(i) *Taller plant species will have less leaf damage by slugs than shorter species.*

Prediction 1A(ii) *Leaf damage will decrease the further up a plant that samples are taken.*

Of course, this is a very broad hypothesis, and many features of a leaf affecting its likelihood of predation are going to change along with its height off the ground. Some of these are suggested by other observations. For example:

Observation Larger leaves are often tougher or smell strongly.

Hypothesis 1B *The decrease in damage among larger leaves is due to their reduced palatability.*

Prediction 1B *For any given size of leaf, damage will decrease the tougher the cuticle or the stronger the odour on crushing.*

Observation Larger leaves are sometimes associated with thorns or sticky hairs on the stems.

Hypothesis 1C *Reduced damage among larger leaves is due to grazing deterrents on the stems.*

Prediction 1C(i) *The incidence of severe damage (suggestive of large herbivores) will be lower on thorny species.*

Prediction 1C(ii) *The incidence of less severe damage (suggestive of invertebrate herbivores) will be lower on species with sticky, hairy stems.*

Example 2
Hosts and parasites

The notes on the samples of material and parasites taken from voles suggest a number of interesting possibilities, some to do with the age and sex of the voles, others to do with relationships between the different parasites.

Observation The number of *Babesia*-infected red cells and faecal egg scores appeared to be higher in male voles than in females, and higher in adults than in juveniles.

Hypothesis 2A *Parasite burdens are affected by differences in the levels of reproductive hormones between age and sex classes.*

Prediction 2A *Parasite burdens will increase with host testosterone levels.*

Hypothesis 2B *Parasite burdens are affected by differences in aggressive behaviour and stress between age and sex classes.*

Prediction 2B(i) *Parasite burdens will be greater among dominant territorial males.*

Prediction 2B(ii) *Parasite burdens will increase in any individual with the amount of aggressive behaviour shown.*

Prediction 2B(iii) *Parasite burdens will increase with host corticosterone (stress hormone) levels.*

Observation Sex and age differences in *Babesia* levels appear to be associated with the number of ticks on the host.

Hypothesis 2C *The intensity of infection with* Babesia *depends on the degree of exposure to infected ticks.*

Prediction 2C Babesia *burden will increase with the number of infected ticks establishing on the host.*

Hypothesis 2D *The intensity of infection with* Babesia *depends on the degree of resistance to tick infection.*

Prediction 2D *The intensity of infection with* Babesia *will decrease with the host's ability to mount an antibody response to ticks.*

Example 3	Observations on the samples of soil-dwelling nematodes suggest a number of things vary with the pollution status of the site of origin – species diversity and fecundity among them.
Nematodes and pollutants	

Observation Fewer species were identified in the samples from the two polluted sites compared with the unpolluted site.

Hypothesis 3A *Pollution reduces species diversity.*

Prediction 3A *The addition of pollutants to identical multi-species nematode cultures will result in a reduction in the number of species supported over time.*

Hypothesis 3A is another very broad hypothesis and could give rise to several more specific hypotheses, each generating its own predictions. For example:

Hypothesis 3B *Pollutants are toxic to those species missing from polluted sites.*

Prediction 3B *Species present at unpolluted sites but missing from polluted sites will show greater mortality when exposed to pollutants.*

Hypothesis 3C *Pollutants affect resource availability for certain groups of nematodes.*

Prediction 3C *Species missing from polluted sites will tend to come from certain trophic or microhabitat groups.*

Observation Some species are present only in polluted sites.

Hypothesis 3D(i) *Such species benefit from relaxed interspecific competition in polluted sites.*

Prediction 3D(i) *Increasing the number of species in the culture in an otherwise constant and pollutant-free environment will tend to result in the loss of such species from the community.*

Hypothesis 3D(ii) *Pollutants create niche opportunities not available in unpolluted sites.*

Prediction 3D(ii) *For any given number of species in the culture in an otherwise constant environment, such species will do better when pollutant is added compared with an unpolluted control.*

Observation Fewer juvenile stages were recorded from the organophosphate-polluted site than from the other two sites, but there was no consistent difference in the number of females with eggs.

Hypothesis 3E(i) *Organophosphate pollution affects recruitment to nematode populations through reduced egg viability.*

Prediction 3E(i) *Females reared in organophosphate-treated, single-species culture will show reduced hatching success per egg compared with those reared in heavy metal or untreated control cultures.*

Hypothesis 3E(ii) *Organophosphate pollution affects recruitment to nematode populations through increased juvenile mortality.*

Prediction 3E(ii) *Females reared in organophosphate-treated, single-species culture will show comparable hatching success per egg but reduced survival of resultant juvenile stages relative to those reared in heavy metal or untreated control cultures.*

| Example 4 |
| Crickets |

Male field crickets seemed to be aggressive to one another when put together in an arena. Whether or not an encounter resulted in fighting varied with the number of crickets, and individuals differed in their tendency to initiate and win fights. The apparent effects of providing egg box shelters and introducing females suggest that interactions between males are concerned ultimately with gaining access to females.

Observation The number of encounters leading to a fight was lower when more crickets were present.

Hypothesis 4A *The cost of fighting on encounter increases with population size and the chance of encountering another male.*

Prediction 4A *The probability of an encounter's resulting in a fight will decrease with increasing numbers of males and in the same number of males maintained at a higher density.*

Observation Larger males initiated more fights per encounter and won in a greater proportion of encounters.

Hypothesis 4B *Large size confers an advantage in fights between males.*

Prediction 4B *Males will be less likely to initiate a fight when their opponent is larger.*

Observation Interactions tended to escalate from chirping and antenna-tapping to overt fighting.

Hypothesis 4C *The escalating sequence reflects information-gathering regarding the size of the opponent and the likelihood of winning.*

Prediction 4C *Encounters will progress further when opponents are more similar in size and it is more difficult to judge which will win.*

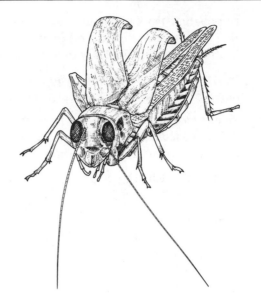

Observation Larger males ended up in or near egg box shelters and females tended to spend more time with these males.

Hypothesis 4D(i) *Females prefer to mate with males in shelters for protection from predators.*

Prediction 4D(i) *Giving a male a shelter will increase the attention paid to him by females and his chances of copulating.*

Hypothesis 4D(ii) *Females prefer large males.*

Prediction 4D(ii) *Given a choice of males, all with or without shelters, females will spend more time and be more likely to copulate with larger males.*

2.3.2 Null hypotheses

In the examples above, we have phrased predictions in terms of the outcomes they lead us to expect. Prediction 1B, for example, leads us to expect that the amount of damage sustained by a leaf will decrease the tougher its cuticle or the more volatiles it contains. We can test this prediction by carrying out a suitable investigation and associated confirmatory analysis. Formally, however, we do not test predictions in this form. Rather, we test them in a null form that is expressed as a hypothesis *against* the prediction. This is known as a *null hypothesis* and is often expressed in shorthand as H_0 (the prediction made by the study is then referred to conventionally as H_1). Predictions are tested in the form of a null hypothesis because science proceeds conservatively, always assuming that something interesting is *not* happening unless convincing evidence suggests, for the moment, that it might be. In the case of Prediction 1B, therefore, the null hypothesis would be that tougher cuticles or more volatiles would make no difference to the amount of damage sustained by a leaf. We shall see later what burden of proof is necessary to enable us to reject the null hypothesis in any particular case.

There is a second, and from a practical point of view more crucial, point to make about the predictions. Skimming down them gives the impression of specificity and diversity; each prediction is tailored to particular organisms and circumstances, and those from one example seem to have little to do with those from others. At the trivial level of detail, this is obviously true. However, in terms of the kinds of question they reflect, predictions from the different examples in fact have a great deal in common. Before we can proceed with the problem of testing hypotheses and choosing confirmatory analyses, we need to be aware of what these common features are.

2.3.3 Differences and trends

Although we derived some 23 different predictions from our notes, and could have derived many more, all fall (and any others would have fallen) into one of two classes. Regardless of whether they are concerned with nematodes or crickets or with surviving pollutants or fighting rivals, they either predict some kind of *difference* or they predict some kind of *trend*. Recognising this distinction is vitally important, because it determines the kind of confirmatory test we shall be looking to perform and therefore the design of our experiments. Surprisingly, however, it proves a stubborn problem for many students throughout their course, with the result that confirmatory analyses often fall at the first fence. Let's look at the distinction more closely.

A *difference* prediction is concerned with some kind of difference between two or more groups of measurements. The groups could be based on any characteristics that can be used to make a clear-cut distinction; obvious examples could be sex (e.g. a difference in body size between males (Group 1) and females (Group 2)), functional anatomy (e.g. a difference in enzyme activity between xylem (Group 1), phloem (Group 2) and parenchyme (Group 3) cells in the stem of a flowering plant), or experimental treatment (e.g. a difference in the number of chromosomal abnormalities following exposure to a mutagen (Group 1) or exposure to a harmless control (Group 2)). Which of the predictions we derived earlier are difference predictions?

Example 1

Plants and herbivores

In the leaf sample study there are two difference predictions:

- Prediction 1C(i) leads us to expect that the incidence of severe damage will be lower on thorny species (Group 1) than on non-thorny species (Group 2).

- Prediction 1C(ii) leads us to expect a reduction in less severe damage on species with sticky, hairy stems (Group 1) compared with species without (Group 2).

Example 2

Hosts and parasites

One prediction from the example of host/parasite relationships involves a difference:

■ Prediction 2B(i) suggests a difference in parasite burden between dominant territorial males (Group 1) and other age and sex categories of host (Group 2).

Example 3

Nematodes and pollutants

Almost all the predictions arising from the soil-dwelling nematode samples turn out to be difference predictions:

■ Prediction 3A suggests a difference in the number of species between cultures to which pollutant has been added (Group 1) and those that are pollutant-free (Group 2).

■ Prediction 3B predicts a difference in sensitivity to pollutants between species absent from polluted sites (Group 1) and those present at such sites (Group 2).

■ Prediction 3C suggests a difference between trophic groups (e.g. bacterial feeders (Group 1), fungal feeders (Group 2), plant feeders (Group 3)) in the tendency to be present at polluted sites.

■ Prediction 3D(ii) leads us to expect a difference between cultures to which pollutant has been added (Group 1) and pollutant-free controls (Group 2) in the tendency to support nematode species found only in polluted sites in the field.

■ Prediction 3E(i) suggests a difference in egg hatching success between female worms exposed to organophosphate pollutant (Group 1) and those not (Group 2).

■ Prediction 3E(ii) is similar to the last prediction except that it suggests a difference in larval mortality instead.

Example 4

Crickets

Two predictions from the crickets involve differences:

■ Prediction 4B suggests that males will be less likely to intitiate a fight when their opponent is larger than them (Group 1) than when it is smaller (Group 2).

■ Prediction 4D(i) predicts that females will pay more attention to males with a shelter (Group 1) than to males without a shelter (Group 2).

Trend predictions are concerned not with differences between hard and fast groupings but with the relationship between two more or less continuously distributed measures. Thus, for example, a relationship might be predicted between the amount of an anthelminthic drug administered to a rat infected with nematodes and the number of worm eggs subsequently counted in the animal's faeces. In this case, we should expect the relationship to be negative with egg counts decreasing the more drug the rat has received. On the other hand, a positive relationship might be predicted between the number of hours of sunlight received and the standing crop of a particular plant. With trends we can therefore envisage two measures as the axes of a graph. One measure extends along the bottom (x) axis, the other up the vertical (y) axis. Sometimes it doesn't matter which measure goes along the x-axis and which up the y-axis because there is no basis for implying cause and effect, and we are interested only in whether there is some kind of association. Thus, we might expect a strong association between the amount of ice cream eaten and the amount of time spent in the sea on a visit to the seaside because both would go up with temperature. Since neither could reasonably be thought of as a cause of the other, it is of no consequence which goes on the x- or y-axis. In the two examples above, however, there are reasonable grounds for supposing cause and effect. While it is plausible for the anthelminthic drug to affect faecal egg counts, it is not plausible for the egg counts to have influenced the amount of drug. Similarly, hours of sunlight could influence a standing crop but not vice versa. In these cases, the drug dose and hours of sunlight measures should go on the x-axis and the egg counts and standing crops on the y-axis. It is important to stress, however, that by doing this we are not asserting that the x-axis measure really *is* a cause of the y-axis measure – as we shall see later, inferring cause and effect from relationships requires extreme caution – merely that if there was a cause-and-effect relationship it would most likely be that way round. This is also clear in the remainder of our example predictions, all of which involve trends.

Example 1 **Plants and herbivores**	In Prediction 1A(i), leaf damage is expected to decrease as the height of plant species increases. Plant height should thus be the x measure and leaf damage the y measure:

- Prediction 1A(ii) makes a similar prediction except that the expected relationship is *within* plants. Height up the plant is the x measure and leaf damage once again the y measure.

- Prediction 1B suggests a negative relationship between toughness of the cuticle (x measure) and leaf damage (y measure).

Example 2 **Hosts and parasites**	All except one of the predictions from the vole parasites example are trend predictions:

- Prediction 2A predicts an increase in parasite burdens (y measure) with increasing testosterone levels (x measure).

- Predictions 2B(ii) and (iii) predict similar increases in parasite burdens (y measure) but this time as a function of increasing aggression and corticosterone levels (x measures), respectively.

- Prediction 2C suggests an increase in *Babesia* levels (*y* measure) with the number of ticks recovered from the host (*x* measure).

- Prediction 2D suggests a reduction in *Babesia* levels (*y* measure) with host immune responsiveness (*x* measure).

Example 3

Nematodes and pollutants

Only one of the predictions arising from the nematode example suggests a trend:

- Prediction 3D(i) predicts a loss of species found only at polluted sites (*y* measure) as the number of species in a culture increases (*x* measure).

Example 4

Crickets

Three predictions from the crickets involve trends:

- Prediction 4A first of all predicts that the number of encounters ending up in a fight (*y*) will increase with the number of crickets (*x*), then predicts a similar increase when the same number of crickets are maintained at higher densities (density = *x*).

- Prediction 4C involves a predicted trend in the tendency to escalate an interaction (*y*) with decreasing difference in size between opponents (*x*).

- Prediction 4D(ii) suggests that the time females spend with a male (*y*) and their tendency to copulate (*y*) will increase with male size (*x*).

There is thus a clear distinction between difference and trend predictions. Of course, it is possible to recast some trend predictions as difference predictions (for instance, a continuous measure of group size for use in a trend could always be recast in terms of small groups (groups below size *w*) and large groups (groups above size *w*) and thus be used in a difference prediction). What makes the distinction, therefore, is not the data per se but the way data are to be collected or classified for analysis. Thus, while measures such as group size or time intuitively suggest trends, there is nothing to stop their being used in difference predictions. It all depends on what is being asked. This is often a source of serious confusion among students encountering open-ended data handling for the first time.

2.4 Summary

1. Open-ended observation is a good way to develop the basis for forming hypotheses and predictions about material. It pays to make observations quantitative where possible so that exploratory analyses can highlight points of interest.

2. Exploratory analysis is a useful (often essential) first step in extracting interesting information from observational notes or other sources of exploratory information. It can take a wide variety of forms, such as bar charts, scattergrams, or tables of summary statistics.

3. Exploratory analyses, or raw exploratory information itself, can lead to a number of hypotheses about the material. In turn, each hypothesis can give rise to several predictions that test it. Formally, predictions are tested in the form of null hypotheses.

4. While predictions derived from hypotheses may be diverse and specific in detail to the material of interest, they fall into two clearly distinguishable categories: predictions about *differences* and predictions about *trends*. Which of these categories a prediction belongs to is determined by the way data are to be collected or classified for analysis.

References

Crawley, M. J. (2007) *The R book*. Wiley, London.

Dytham, C. (2010) *Choosing and using statistics: A biologist's guide*. 3rd edition, Blackwell, Oxford.

3 Answering questions
What do the results say?

In the last chapter, we looked at the way hypotheses can be derived from exploratory information. We turn now to the problem of how to test our hypotheses. As we have seen, we begin by making predictions about what should be the case if our hypotheses are true. These predictions then dictate the experiments or observations that are required. However, this may not be as straightforward as it sounds; decisions have to be made about what is to be measured and how, and how the resulting data are to be analysed. The questions of measurement and analysis are, of course, interdependent. This is obvious both at the level of choosing between difference and trend analyses – there is little point collecting data suitable for a difference analysis if what we're looking for is a trend – and at the choice of analyses *within* differences and trends. While at first sight it might seem like putting the cart before the horse, therefore, we shall introduce confirmatory analysis *before* dealing with the collection of data so that the important influence of choice of analysis on data collection can be made clear.

3.1 Confirmatory analysis

3.1.1 The need for a yardstick in confirmatory analysis: statistical significance

Take a look at the scattergram in Fig. 3.1. It shows a relationship between the concentration of a fungicide sprayed on a potato crop and the percentage of leaves sampled subsequently that showed evidence of fungal infection. A plot like this was presented to a class of undergraduates. Students in the class were

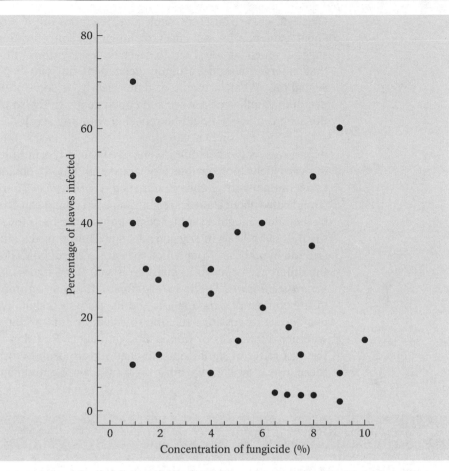

Figure 3.1 A scattergram of the relationship between concentration of a fungicide applied to a potato crop and the percentage of leaves subsequently found to be infected.

asked whether they thought it suggested any effect of fungicide concentration on infection. The following are some of their replies:

Yes, fungicide concentration obviously has an effect because infection goes down with increasing concentration.

It's hard to say. It looks as though there is some effect, but it's pretty weak. More data are needed.

Fungicide concentration is reducing infection but there must be other things affecting it as well because there's so much scatter.

I don't think you can say anything from this. Yes, there is some downward trend with increasing concentration but several points for high concentrations are higher than some of those for low concentrations. Totally inconclusive.

Yes, there is a clear negative effect.

Clearly there are different, subjective, reactions to the plot. To some it is unequivocal evidence for an effect of fungicide concentration; to others it doesn't suggest much at all. Left to 'eyeball' impressions, therefore, the conclusion that emerged would be highly dependent on who happened to be doing the eyeballing. What is required, quite clearly, is some independent yardstick for deciding whether or not we can conclude anything from the relationship. Since the scenario above could be repeated with any set of data – difference or trend – the need for such a yardstick arises in all cases. One or two idiosyncratic departures notwithstanding (one well-known ornithologist used to advocate the yardsticks 'not obvious', 'obvious' and 'bloody obvious'), the yardstick used conventionally in science is *statistical significance*. There is nothing magical or complicated about statistical significance. It is simply an arbitrary criterion accepted by the international scientific community as the basis for rejecting or not rejecting the null hypothesis in any given instance and thus deciding whether predictions, and the hypotheses from which they are derived, hold. If the criterion is reached, the difference or trend in question is said to be *significant*; if it is not, the result is *non-significant*. The term 'significant' thus has an important, formal meaning in the context of data analysis and its use in a casual, everyday sense should be avoided in discussions relating to scientific interpretation. How do we decide whether differences or trends are significant? By using the most appropriate of the vast range of significance tests at our disposal. Before we introduce some of these tests, however, we must say a little more about significance itself.

3.2 What is statistical significance?

A *statistic* is a measure, such as a mean or a correlation, derived from samples of data that we have collected. Our expectation is that it relates closely to an equivalent, real value (*parameter*) in the population from which the samples were drawn. Of course it might or it might not. Our sampling technique (*see later*) may have been impeccable and produced a very accurate reflection of the real world. More than likely, however, and for all kinds of forgivable reasons, it will have produced a somewhat biased sample and our calculated statistics will differ from their population parameters. This is what makes statistical inference tricky. If we detect an apparent difference between two sets of data, or an apparent trend in the relationship between them, is the difference or trend real or is it just an artefact of the chance bias in our sampling? Or, to put it another way, is it statistically significant?

The criterion that determines significance is the probability (usually denoted as p) that a difference or trend as extreme as the one observed would have occurred if the null hypothesis – that there is really *no* difference or trend in the population from which the sample came – was true. Confused? An example makes it clear. Let's take one of our earlier predictions, say Prediction 2C. This predicts an increase in *Babesia* burden in voles with increasing degree of tick infestation. Suppose we had tested this by infesting each of ten sets of five parasite-free voles with a different number of *Babesia*-infected ticks, measured the subsequent number of infected blood cells in each vole and found what looked like a convincing positive trend: *Babesia* burden goes up with increasing numbers of ticks. The null hypothesis in this case, of course, is that *Babesia* burden will *not* increase with the number of ticks. What, then, is the probability of obtaining a positive relationship as extreme as the one we got if this null hypothesis is really true and the apparent trend a chance effect? A helpful analogy here might be the probability of obtaining the apparent trend by haphazardly throwing darts at the scattergram. An appropriate significance test will tell us (we shall see how later). By convention in biology, a probability of 5 per cent (= 0.05 when expressed as a proportion) is accepted as the threshold of significance. If the probability of obtaining a relationship as extreme as ours by chance turns out to be 5 per cent or less, we can regard the relationship as significant and reject the null hypothesis. If the probability is greater than 5 per cent we do not reject the null hypothesis, and the relationship is regarded as non-significant. If the null hypothesis is not rejected, we effectively assume that our apparent relationship was due to a chance sampling effect. As a matter of interest, the negative trend in Fig. 3.1 is significant at the 5 per cent level, so the optimists have it in this case!

The 5 per cent threshold is, of course, arbitrary and still leaves us with a one-in-twenty chance of rejecting the null hypothesis incorrectly (falsely accepting there is a difference or a trend when there isn't). Under some circumstances, for instance when testing the effectiveness of a drug, a one-in-twenty risk of incorrect rejection might be considered too high. In certain areas of research such as medicine, therefore, the arbitrary threshold of significance is set at 1 per cent (= 0.01). In other disciplines it is sometimes relaxed to 10 per cent (= 0.1). Although we have talked of threshold probability (p) values ($p < 0.05$, $p < 0.01$, etc.), most computer statistical packages now quote *exact* probabilities ($p = 0.0425$, $p = 0.1024$, etc.) for the outcome of significance tests. If the package doesn't tell you whether the exact probability it quotes is significant, simply apply the threshold value rule as before. Thus, on the 5 per cent criterion, $p = 0.0425$ is significant, because it is less than 0.05, but $p = 0.1024$ is not, because it is greater than 0.05[†]. Many students are thrown by the appearance of a probability of 0.000 from a statistical package. Don't be! It merely means a probability too low to display within three decimal figures; it therefore means $p \ll 0.001$ (p is very much less than 0.001).

[†] There are differing opinions about the form in which the p-value from a statistical test should be reported. Some take the view that, since most packages give the exact probability, this should be reported as the result. However, some statisticians think that these probabilities are really correct only if the assumptions of the test are absolutely fulfilled by the data, which is rarely true, and hence it is misleading to give the appearance of accuracy by citing the exact probability. For many scientists the threshold is the important thing, and they report only whether (a) it has not been equalled or crossed (in which case the outcome is reported as 'non-significant' or 'ns'), or (b) if it has been crossed, at what level of probability, either by quoting the thresholds themselves ($p < 0.05$, $p < 0.01$ or $p < 0.001$) or using a conventional indicator such as asterisks ('*', '**' or '***') to denote the same thing (*see also* p. 86). However, see the interesting paper by Ridley *et al.* (2007), which suggests scientists sometimes pursue these critical thresholds rather too zealously.

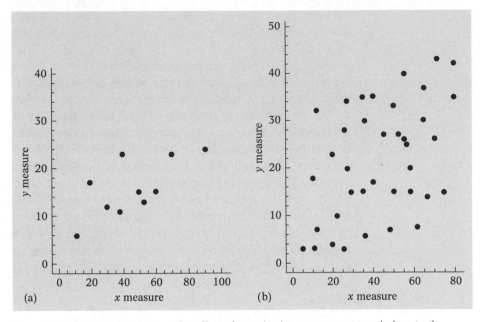

Figure 3.2 Scattergrams showing the effect of sample size on apparent trends (*see text*).

An important point must be made here regarding the inference to be drawn from achieving different levels (10, 5, 1 per cent, etc.) of significance. A high level of significance is not the same as a large effect in the sense of a large difference or a steep trend. The magnitude of an effect – difference or trend – is usually known as an *effect size*[†]. This is quite different from the level of significance. The limitations of significance in this sense are made clear in Fig. 3.2. The figure shows two trends. In Fig. 3.2a, the *y* measure increases in close relationship with the *x* measure, yielding what looks like a clear positive trend. Figure 3.2b, on the other hand, shows a scatter of points in which it is more difficult to discern a trend. However, if we perform a suitable significance test for a trend on the two sets of data, the trend in Fig. 3.2b turns out to be significant at the 1 per cent level while that in Fig. 3.2a isn't even significant at the 10 per cent level. The crucial difference between the two trends, of course, is the sample size. The number of data points in Fig. 3.2a is low, so a few inconsistencies in the trend are enough to push it below significance. Figure 3.2b, however, has a large number of points; so even though there is a wide scatter, the trend is still significant. Exactly the same-sample size effect would operate in the case of difference analyses.

Because the level of significance by itself gives little indication of the magnitude of a difference or trend, it is always important to provide such an indication, usually in the form of summary statistics and their associated sample sizes. We shall return to this point later.

[†] 'Effect size' is a name given to a family of indices that measure the magnitude of a difference or trend. Unlike significance tests, effect sizes are independent of sample size. They are usually measured in two ways: (a) as the standardised difference between means (difference analyses) or (b) as the correlation between the predictor variable and the response variable (trend analysis) (*see also* Box 3.10).

3.3 Significance tests

So far, we have seen how to get round the problem of subjective impression in interpreting data by using the criterion of statistical significance, and we have looked at some caveats on the interpretation of significance. We can now turn to the statistical *tests* that enable us to decide significance.

Right at the beginning we said that this book was not about statistics. It isn't. At the same time, statistical significance tests are an essential tool in scientific analysis, and the rules for using them must be clearly understood. This does not necessarily require a knowledge of statistical theory and the mathematical mechanics of tests any more than using a computer program requires an appreciation of electronics and microcircuitry. As with most tools, it is competent *use* that counts rather than theoretical understanding. But while our aim here is simply to introduce the use of significance tests as a basic tool of enquiry, we stress again (*see* Box 2.2) that acquaintance with statistical theory is strongly recommended, and it is envisaged that many users of this book will also be pursuing courses in statistics and have at their disposal some of the many introductory and higher-level textbooks now available (e.g. Siegel & Castellan, 1988; Sokal & Rohlf, 1995; Grafen & Hails, 2002; Dytham, 2010; Hawkins, 2005; Ruxton & Colegrave, 2006; Crawley, 2007; Zuur *et al.*, 2009).

3.3.1 Types of measurement and types of test

The test we choose in a particular case may depend on a number of things. The following are three important ones.

1 Types of measurement
The first is the kind of measurement we employ. Without getting too bogged down in jargon, we can recognise three kinds.

Nominal or classificatory measurement. Here, observations or recordings are allocated to one of a number of discrete, mutually exclusive categories such as male/female, mature/immature, red/yellow/green/blue, etc. Thus if we were to watch chicks pecking at red (R), green (G) and orange (O) grains of rice and recorded the sequence of pecks with respect to the colour targeted, we might end up with a string of data as follows:

R O R R G G O R G G G O O G R O

Such data are measured purely at the level of the category to which they belong, and measurement is thus *nominal* or *classificatory*.

Ordinal or ranking measurement. In some cases, it may be desirable (or necessary) to make measurements that can be *ranked* along some kind of scale. For instance, the intensity of the colour of a turkey's wattles might be used as a guide to its state of health. The degree of redness of the wattles of different birds could be scored on a scale of 1 (pale pink) to 10 (deep red). The allocation of scores to wattles is arbitrary and there is no reason to suppose that the degree of redness increases by the same amount with each increase in score. Thus the difference in redness between scores of 8 and 9 might be greater than the difference between scores of 2 and 3. All that matters is that 9 is redder than 8 and 3 is redder than 2; the absolute difference between them cannot be quantified meaningfully.

Constant-interval measurements. In other scale measurements, the difference between scores can be quantified so that the difference between scores of 2 and 3 is the same as that between scores of 8 and 9. Such measurements may have arbitrarily set (e.g. scales of temperature) or true (e.g. scales of time, weight, length) zero points. Such *constant-interval measurements* can in fact be split into two categories on the basis of arbitrary versus true zero points and their scaling properties (e.g. Martin & Bateson, 1993), but this is not important here.

While defining measurements seems rather dry and theoretical, we need to be aware of the kind of measurement we use because, as we shall see, some significance tests are very restrictive about the form of data they can accept. Another reason for highlighting it is that we should always seek the measures that give us the maximum amount of information for the cost required (usually in time) to obtain them. Usually this means constant-interval measurements, because they are on a continuous, non-arbitrary scale. However, nominal or ordinal data are sometimes more appropriate and not infrequently the best that can be achieved.

2 Parametric and non-parametric significance tests

The second thing we must keep an eye on is the nature of the data set itself, in particular the sample size and the distribution of values within the sample. Again, detailed consideration of this is unnecessary, but it is a factor that determines the range of tests we shall be introducing, so a brief discussion is warranted.

Parametric tests. These make a number of important assumptions that are frequently violated by the kinds of data sets collected during practical exercises. The most critical concerns the distribution of values within samples. Parametric tests generally assume that the data conform (reasonably closely at least) to what is known as a *normal* distribution.

As Dytham (2010) puts it, the normal distribution is the most important distribution in statistics (*but see* Box 3.1a) and it is often assumed (all too frequently without checking) that data are distributed in this way. We've already encountered it in our discussion of frequency distributions and it is illustrated again in Fig. 3.3. Essentially a normal distribution demands that most of the data values fall in the middle of the range (cluster about the mean) with the number tapering off symmetrically either side of the mean to a few extreme values in each of the two tails. The height of the adult male or female population in a city would look something like this: most people would be around the average height for their sex, some would be quite tall or quite short and a few would be extremely tall or extremely short. While normality is not the only assumption underlying parametric tests, the arithmetic of such tests is based on the parameters describing this symmetrical, bell-shaped curve (hence the term *parametric*). Therefore,

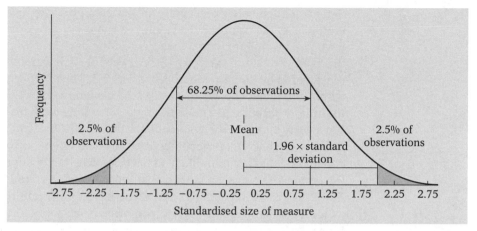

Figure 3.3 A standardised normal distribution (with a mean of 0) and showing one standard deviation (vertical lines) either side of the mean (modified from Dytham, 2010).

if you are testing for group differences, parametric tests assume that each group has the same normal distribution of values around its mean; if for trends, that there is a normal distribution of y-values at each x-value (for regression), or a normal distribution along both x- and y-axes (for correlation). The more distorted (less normal) the distribution becomes, therefore, the less meaning the calculations of parametric tests have. We can convince ourselves of this by briefly considering some basic features of the distribution. We shall then consider what to do if the data are not normal.

The standard deviation and probability. We've talked about the arithmetic mean as a measure of central tendency. This is a useful parameter that tells us one thing about the nature of the data set. What it doesn't tell us, of course, is anything about the *variation* in the data. This is where we come back to the *standard deviation*, first encountered in calculating the standard error to the mean in Box 2.1. The standard deviation (*see* calculation in Box 2.1) is a measure of the spread of data values about the mean, but instead of simply being the full range of *actual* values from smallest to largest (which will vary with sample size), it reflects the theoretical spread of the majority of values (68.25 per cent of them in a perfectly normal distribution) in the true population. We can represent this as two vertical lines (one either side of the mean) in Fig. 3.3, which is actually a standardised normal distribution (where the mean is subtracted from each data value and the result divided by the standard deviation, thus giving a mean of zero and a variance of 1). An easy way to visualise the standard deviation is as the point of inflection either side of the mean (where the curve of the normal distribution changes from convex to concave). Taking the majority spread like this avoids the distortion that might be inflicted by odd outlier values at the extremes, and provides a convenient standard yardstick of variability within data sets. Moving two or three standard deviations away from the mean includes predictably greater percentages of the data set: 94.45 per cent in the case of two standard deviations and 99.97 per cent in the case of three. The important point here is that we can use this property when it comes to significance testing.

Since we know what percentage of the data points is included within different multiples of the standard deviation, we can easily work out how many standard deviations would include 95 per cent of the data or, more to the point, exclude the 5 per cent at the extremes (2.5 per cent at each extreme). The answer is 1.96, represented by the horizontal bar in Fig. 3.3. Thus 95 per cent of data values lie within the range 'mean ± 1.96 standard deviations ($\bar{x} \pm 1.96$ s.d.)', which means that the probability of encountering a value larger or smaller than this range is 5 per cent or less – the conventional threshold of statistical significance! If we found a value with a probability of occurrence as low as this we'd be justified in concluding that it was unlikely to have come from the population that generated the curve. Of course, as the fact that the calculation of the standard deviation in Box 2.1 is based on squared deviations makes clear, all this works *only* as long as the curve is symmetrical and conforms respectably to a normal distribution. Since the calculation of parametric significance tests generally employs the same squared deviations procedure, as we shall see shortly, the restrictions apply to all these tests.

One additional point. Quite a number of significance tests are based on calculating the deviation of an observed mean from the null expectation of a standardised normal distribution (Fig. 3.3). Conventionally this is called a z-value and, from the above, we can see that if z exceeds 1.96 the result is significant at the 5 per cent level. (The value 1.96 is actually the threshold value when both sides – 'tails' in the jargon – of the distribution are taken into account, a so-called two-tailed test; when only one tail is considered (a one-tailed test), the threshold value is 1.64. We shall discuss one- and two-tailed tests in detail later.)

Departures from normality. Frequency distributions can depart from normality in a number of ways. Two broad kinds of departure, however, are *skewness* and *kurtosis*. Skewness is a synonym for asymmetry, i.e. one or other tail of the distribution is more drawn out than the other. The distribution in Fig. 2.8b is said to be skewed to the right (towards the *y*-axis), while the opposite bias would be skewed to the left. Kurtosis refers to the flatness of the distribution, which can be *leptokurtic* (more values are concentrated around the mean and in the tails and fewer in the 'shoulders' of the distribution), or *platykurtic* (where the reverse is true). Bimodal distributions, like that in Fig. 2.8c, are thus extremely platykurtic.

Percentages and proportions present their own problems for normality. Because they range between 0 and 100, or 0 and 1, the distribution of values is artificially truncated at either end. This may not present a serious problem if most of the values in the data set occur in the middle two-thirds or so of the distribution, but if they approach 0 or 100/1 there is cause for concern.

Transformations. So what do we do if we suspect our data may not be normally distributed? Happily, and as long as our sample size is big enough (> 50 as a rough guide) to make a comparison meaningful, the wide range of statistical packages now available for personal computers makes the answer simple. Test it! There are some well-established significance tests (such as the Shapiro–Wilk and Kolmogorov–Smirnov tests) that allow comparisons between frequency distributions of data and various theoretical distributions, of which the normal is the commonest. But there is a catch, and Box 3.1b explains this in showing how to test for normality using R. If we test our distribution and find it does not differ significantly from normal, then we're at liberty to use any appropriate parametric test at our disposal. If it does differ, we can do one of two things. We can abandon the idea of using parametric statistics and choose an appropriate *non-parametric* test instead (*see below*), or we can *transform* the data to see whether they can be normalised. Several transformations are available but the most widely used are probably *logarithmic* (log *x* or, where there are zeros in the untransformed data, $\log(x + 1)$) or *square root* transformations. Simply log or take the square root of each data value, recast the distribution and test it for normality again. Where percentages or proportions stray below about 30 per cent (or 0.3) or above 70 per cent (or 0.7), an *arcsine square root* transformation (calculated by taking the square root of the proportion – so divide percentages by 100 first – then the inverse sine ($\sin^{-1}$ on many calculators) of the result) will stretch out the truncated tails and prevent undue violation of the normality assumption of parametric tests (Box 3.1b shows how to transform data in R).

Of course, even transformation may not succeed in normalising our data, in which case we must seriously consider using non-parametric statistics. Indeed, we may not even get as far as worrying about the normality of our data before opting for a non-parametric approach. Among their various other requirements, parametric tests demand that measurements are of the constant-interval kind, so cannot usually deal with the other types of measurement we might be forced to use. Non-parametric tests are much less restrictive here.

Non-parametric tests. Non-parametric tests are sometimes referred to as distribution-free, *ranked* or *ranking* tests because they do not rely on data being distributed normally, and generally work on the ranks of the data values rather than the data values themselves. While they may be distribution-free, however, they are not entirely assumption-free. They assume the data have *some* basic properties, such as independence of measurement (*see later*) and a degree of underlying continuity (*see* Martin & Bateson, 1993): crucially, they also make the same assumption of equal dispersion among groups as do parametric tests (which assume equal variances among groups). In most cases, however, these assumptions are easily met. In the jargon of statisticians, non-parametric tests are thus more *robust* because they are capable of dealing with a much wider range of data sets than their parametric equivalents. While they can deal with the same constant-interval measurements as parametric tests, they can also cope with ordinal (ranking) and nominal (classificatory) measurements. Non-parametric tests are especially useful when sample sizes are small and assumptions about underlying normality particularly troublesome. There are a couple of drawbacks, however. The first, arguably overstated (*see* Martin & Bateson, 1993), weakness is that non-parametric tests are generally slightly less powerful (*power* here meaning the probability of properly rejecting the null hypothesis – we shall return to this shortly) than their parametric equivalents. The second, which is slowly being addressed (*see for example* Meddis, 1984), is that the range of

Box 3.1a   Types of distribution

The normal (or Gaussian) distribution is a form of *continuous* distribution, where, in principle, data can take a continuum of values. There are other forms of continuous distribution, but the normal is by far the most relevant to biological situations. However, there is another family of distributions which is important in biology. These are *discontinuous* or *discrete* distributions: three are particularly relevant.

The Poisson distribution

The Poisson distribution describes occurrence in units of time or space, for instance the number of solitary bee burrows scored in a quadrat or the number of hedgehogs dropping into a cattlegrid overnight. The key assumptions are: (a) the mean number of occurrences per unit is small in relation to the maximum number possible – i.e. occurrences are rare, (b) occurrences are independent of each other – i.e. there is no influence of one on the likelihood of another, (c) occurrences are random. Indeed, the reason a Poisson distribution is fitted to data is to test for independence or randomness in time or space. Poisson data are characterised by the mean being equal to the square of the standard deviation (known as the *variance*). If the variance is bigger than the mean, the data are more clustered than random; if it is smaller, they tend towards uniformity. A glance at these two summary statistics thus gives a good indication of the kind of distribution we're dealing with. When analysing count data with small numbers close to zero, the Poisson distribution also makes the crucial assumption that no values can be *below* zero. It can be very misleading to use statistical tests based on the normal distribution here, because this assumes the data can take any value, including negative ones. It is only relatively recently that statistical methods have been available that can make the assumption that the data are Poisson-distributed. The ability to use these and the other distributions on this page in Generalised Linear Models is a great step forward, reducing the need for non-parametric methods in many instances.

The binomial distribution

The binomial distribution is a discrete distribution of occurrences where there are two possible outcomes for each so that the probability of one outcome determines the probability of the other. Thus if the probability of an egg's hatching is 0.75, the probability that it won't hatch is 1 – 0.75, i.e. 0.25. This logic can be extended to calculate the probability of various combinations of hatching (H) and failed (F) eggs in clutches of different size. Thus if two eggs are laid, there are four possible outcomes: HH, HF, FH, FF. Applying the values above gives a probability of 0.56 (0.75 × 0.75) for both eggs hatching successfully, 0.19 (0.75 × 0.25) for each of the two mixed outcomes (HF and FH) and 0.06 for two failures. The same could be done for clutches of three, four or however many eggs. The binomial distribution thus gives a baseline chance probability against which outcomes in various situations can be judged. So, if we found a population in which the proportion of two-egg clutches failing completely was 30 per cent instead of 6 per cent, we might become suspicious about the health of the birds or their environment. As with the Poisson distribution, it makes little sense to analyse probabilities assuming an underlying normal distribution if the data are distributed in a binomial fashion, since values cannot go below zero or above 1. Two kinds of response variables can be analysed in this way: 0/1 data, or proportions where replicates consist of a set of trials in each of which there were N successes (however defined) and M failures.

The negative binomial distribution

This is used where the data are highly aggregated, which is quite a common situation in biology. For example, if we count the number of gut parasites in each individual of a set of male and female voles trapped from the wild, some individuals will contain large numbers while most will have few or none. This is the typical situation where a negative binomial distribution is needed.

Box 3.1b Testing whether data conform to a normal distribution

Given the importance of the normal distribution to parametric statistical analyses, you might be forgiven for imagining that testing whether your data conform to it would be made easy. Actually it isn't straightforward or automatic in any package: you have to understand what it is you are trying to test, and hence how to go about it.

The key point to understand is that all statistical tests involve fitting a *model*. If you are testing for differences, this model usually involves the fitting of mean values to each group, and testing whether these mean values are different among the groups. If you are testing for a trend, then the model usually involves fitting a straight line to the data, and then testing whether the slope of the line is different from zero.

The assumption of normality in these statistical tests is about the *residuals*, *after* you & fitted the model (*see* discussion in Sokal & Rohlf, 1995). A residual is the difference between an individual measurement and the mean value for its group (for analyses of differences between groups), or from the value predicted by a regression line (for analyses of trends). Thus usually you are not testing whether the raw data are normally distributed, but whether the residuals are.[†] The process therefore usually involves (a) fitting the model, (b) saving the residuals, and (c) carrying out *one single* test to see whether the residuals are normally distributed.

Once we have the residuals, a variety of tests can be used to detect departures from a normal distribution. Those used most commonly are the one-sample Kolmogorov–Smirnov test (for large samples, where $n > 2000$), chi-squared and, for

small- to medium-sized samples where n lies between 3 and 2000, the Shapiro–Wilk test. R makes the process of checking assumptions particularly easy.

Testing for normality as part of a test for differences

Assume that we are interested in whether the mean sizes of seven groups of grasshoppers are different, and that we have 100 measurements for each group. One of the assumptions of the test is that the residuals, the differences of each value from the mean for its group, are normally distributed. Another assumption is that the residuals of each group all have the same normal distribution (i.e. the variances of the groups all have the same value – referred to as the 'homogeneity of variances' assumption): we shall show how to test for this assumption, along with the details of exactly how to carry out tests of difference, later on (*see* Box 3.3a on p. 73). For now we are concerned simply with testing for normality.

In R, as with nearly all statistical packages, the data for the response variable ('size') are in a single column, and a second column ('grassh') indexes the group to which each value belongs (here running from 'A' to 'G'). Such group-membership variables are called *factors* or, more generally, *predictors*. Notice that factors are nominal variables. Here the values are alphanumeric so R assumes the variable is a nominal factor: hence we do not need to declare it as such before running the model.

We now 'fit the model' by running a General Linear Model, or glm, holding it in the object 'm1':

```
> m1 <- glm(size ~ grassh)
```

We will see this standard test for a difference among groups later in detail (Box 3.3a). If not asked, R produces no output, so we are not bothered by details of results from the test of differences. However, R holds all the required details in memory that we need for testing for normality. In particular, the residuals from the fitted model

[†] We can understand intuitively why testing the raw data for normality would be a mistake. Suppose, for example, that the raw data reflect two widely separated groups of values, perhaps body height from males and females where males are much taller than females. Since the two groups differ significantly in mean value, the distribution of the overall raw data will be bimodal, i.e. having two peaks of value with a trough in between. This will fail any test for normality, even though the assumption of normality is almost certainly correct (height is the classic case of the normal distribution).

are held in `resid(model)`. To test for normality, we should do three checks: (i) a statistical test of normality; (ii) visually plot the residuals together with a normal distribution; and (iii) look at the quantile–quantile plot (or q–q plot), which plots the ranked residuals against a similar number of ranks produced from a normal distribution – the result should be a straight line, with systematic deviations from this interpretable as various kinds of non-normality. If there is evidence of skew, then we can test for that as well.

For (i), the first check, we simply type/paste:

```
> shapiro.test(resid(m1))
```

and R gives us the result:

```
Shapiro-Wilk normality test
data: resid(m1)
W = 0.9974, p-value = 0.3316
```

Here the null hypothesis is that the distribution is normal, and since the probability is not less than 0.05, we cannot reject this on the basis of these data. This looks fine.

We can also easily do (ii), visualising the distribution of the residuals together with the expected normal distribution. First we plot the histogram of residuals, here running from –6 to +6 with a bar width of 0.5:

```
> hist(resid(m1),
  breaks=seq(-6,
  6,0.5))
```

Then we construct a set of x-values running from –6 to +6.

```
> xv <- seq(-6,6,0.1)
```

and then generate the normal curve for a mean of zero and the observed standard deviation of our residual, using the probability density function `dnorm`. The height of our frequency distribution depends on how many data points there are, so we have to add a scaling factor. A rough guide to the correct scaling factor is the number of data points multiplied by the chosen bar width:

```
> hist.ht <- length(resid(m1))*0.5
> yv <- dnorm(xv,mean=0.0,
  sd=sqrt(var(resid(m1))))*hist.ht
```

and add this to the plot:

```
> lines(xv,yv)
```

From Fig. (i) we see that the fit is really pretty good.

Obtaining (iii), the q–q plot, involves plotting the data (`qqnorm`) and then the line (`qqline`), which is a dashed (`lty=2`) rather than a solid line (`lty=1`). Thus we type:

```
> qqnorm(resid(m1))
> qqline(resid(m1),lty=2)
```

In Fig. (ii) we can see a very good straight line apart from some minor deviations right at the ends. Thus all seems well.

Figure (i) Distribution of the residuals from the model: clearly they are close to normal.

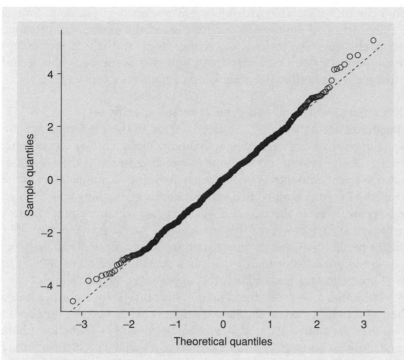

Figure (ii) The *q–q* plot for the mode, showing the straight line expected for the normal distribution.

Testing whether the normal distribution is the same in every group

A very important assumption of parametric tests (i.e. those based on a particular distribution of residuals, usually the normal) is that of constant variance, i.e. that the normal distribution is the same. There are two possible tests for this, Bartlett's and the Fligner–Killeen:

```
> bartlett.test(size ~ grassh)
```

```
Bartlett test of homogeneity of
variances
data: size and grassh
Bartlett's K-squared = 4.7749,
df = 6, p-value = 0.573
```

or

```
> fligner.test(size~grassh)
```

```
Fligner-Killeen test of homogeneity
of variances
```

```
data: size by grassh
Fligner-Killeen:med
chi-squared = 4.4846,
df = 6, p-value =
0.6114
```

Either way we get the same result, which is that there is no evidence to reject the null hypothesis of homogeneity of variances among these groups.

A quick way of producing a set of four diagnostic graphs to help in assessing the assumptions of our test (or 'model') is to fit the model and then use 'plot(model)':

```
> m1 <- glm(size ~
  grassh)
```

```
> plot(m1)
```

and R produces a set of four diagnostic graphs. One is the q–q plot. Others help us to assess outliers and the assumption of homogeneity of variance. It is a good idea to use this routinely whenever we fit any kind of model (see Box 3.15). In fact, R standardises the residuals automatically in different ways according to the assumed distribution of the residuals (normal, Poisson, binomial, etc.), and hence the linearity of the q–q plot is always an important part of checking your model assumptions.

Testing for normality as part of a test for a trend

In R, the way you test for a trend is very similar to the way you test for a difference. The only distinction lies in the predictor, which is a continuous or constant interval variable, rather than a factor. The same set of residuals are produced, which you can test for normality in an exactly similar manner. You can also test for homogeneity of variances in exactly the same way.

tests for more complex analyses involving several variables at the same time is very limited. Sophisticated multivariate analysis is still the undisputed province of parametric statistics. Nevertheless, for our purposes, and with a few exceptions, there are perfectly good parametric and non-parametric equivalents, and we shall introduce them both in our discussion of significance testing.

3 One-tailed versus two-tailed, and general versus specific tests

The third important factor we must consider relates to the prediction we are trying to test. Suppose we are predicting a difference between two sets of data, say a difference in the rate of growth of a bacterial culture on agar medium containing two different nutrients. We could make two kinds of prediction. On the one hand, we could predict a difference without implying anything about which culture should grow faster. In this case, we wouldn't care whether culture A grew faster than culture B or vice versa. This is a *general* prediction. On the other hand, we might predict that one particular culture would grow faster than the other, e.g. A would grow faster than B; this is a *specific* prediction. Which of these kinds of prediction we make affects the way we test the predictions.

The same distinction arises with trend predictions. Imagine we want to know whether there is a trend between the size of a male cricket and the number of fights he wins over the course of a day. We can make a general prediction (there will be a trend, positive or negative), or we can make a specific prediction (larger males will win more fights; i.e. the trend will be positive). We can think of the general prediction as incorporating both positive and negative trends; either would be interesting. The specific prediction is concerned with only one of these.

In cases like those above, where there are only two possible specific predictions within the general one, we can use the same significance test for either general or specific predictions, but with different threshold probability levels for the test statistic (*see below*) to be significant. Because here the specific and general

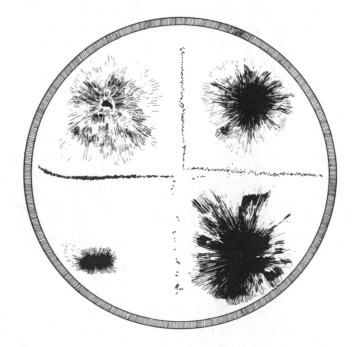

predictions are concerned with one and two directions of effect respectively, the threshold value of the test statistic at the 5 per cent level in the general test becomes the threshold value at the 10 per cent level in the specific test. In statisticians' jargon, we thus do either a one-tailed (specific) or a two-tailed (general) version of the same test. There is, of course, an obvious, and dangerous, trap for the unwary here. The trap is this: if the value of a test statistic just fails to meet the 5 per cent threshold in a two-tailed test, there is a sore temptation to pretend the test is really one-tailed so that the test statistic becomes significant. *It must be stressed that this is tantamount to cheating. A one-tailed test is legitimate **only** when the prediction is made **in advance** of obtaining the result **and** when results in the opposite direction can reasonably be regarded as equivalent to no difference or trend at all[†]. It is completely inadmissible as a fallback when a two-tailed test fails to yield a significant outcome.* A one-tailed test should thus be used only when there are genuine reasons for predicting the direction of a difference or trend *in advance*.

However, the distinction between general and specific is not simply that between two-tailed and one-tailed tests. The latter distinction is normally used for two-group tests such as *t*-tests, where there are clearly only two possibilities for a one-tailed test (A > B or B > A) that together make up the two-tailed case. But the world of specific testing is much richer than this!

There are in fact two different ways of making specific hypotheses, and the distinction is important. You can either predict the *rank order* of the means (values or ranks) of *all* the groups in your experiment, or you can pick out *particular* sets of groups to make a *contrast*. The testing of these two kinds of specific hypothesis differs.

Imagine, for instance, that we are predicting a difference between three (or more) groups – say the weight of fruits produced in a season from trees given three different maintenance treatments: A, B and C. The general prediction is that there will be differences between the three treatments, i.e. that A ≠ B ≠ C. Unlike in the two-group case, this general prediction is 'made up of' *six* potential specific predictions about the *rank order* of the treatment means:

A > B > C, B > C > A, C > A > B, A > C > B, B > A > C and C > B > A

As long as we make the specific prediction *before* we collect the data, we can test any *one* of these in our experiment. Thus, as generally understood, the one-tailed/two-tailed distinction is only a special, and limited, case of the difference between general and specific tests.

There is an alternative way of making specific predictions that uses *contrasts*. These are ways of dividing up ('decomposing' in statistical parlance) the differences between groups into particular patterns. Each contrast consists of one subset of the groups (α) contrasted against a different subset of groups (β),

---

[†] If you predict from theory that A will have a greater mean value than B (i.e. H_1 is A > B), you are also assuming that both of the alternative results (A = B and A < B) are *equivalent* and *together* form the null hypothesis, H_0. Thus if you find that, contrary to your prediction, the mean value of B is much greater than that of A, and would have been significant had you framed your hypothesis as a general one (H_1 is A ≠ B), you are not allowed to conclude *anything* other than that the null hypothesis has *not* been rejected.

effectively creating *two* groups out of the data and comparing them using a two-group test such as a *t*-test (*see below*). A contrast can be general ($\alpha \neq \beta$) or specific (e.g. $\alpha > \beta$), but must be made a priori, i.e. in advance of collecting the data. However, each contrast should be independent of the others, so the general rule is a maximum of one fewer contrasts than the number of groups that could potentially be compared. In the example of the effect of three treatments on fruit production, for example, we could make two independent contrasts (because there are three groups): one of A versus B + C, and the other of B versus C. We shall make use of this approach later (*see*, for example, Boxes 3.9 b, d).

3.3.2 Simple significance tests for differences and trends

Having discussed the general principle of statistical significance, we come now to some actual tests that allow us to see whether differences or trends are significant at an appropriate level of probability. A glance at any comprehensive statistics textbook will reveal a plethora of significance tests for both kinds of analysis. These cater for the various subtleties of assumption and requirement for statistical power under different circumstances. Many of these tests, however, are sophistications of more basic tests that are suitable for a wide range of analyses. Here, we introduce a selection of such tests that can be used with most kinds of data that are likely to be collected in practical exercises. Where appropriate, parametric and non-parametric equivalents are presented. There are some extremely useful non-parametric tests of difference, however, that have been developed independently (Meddis, 1984) and are not available in any existing statistical package as far as we are aware. We have therefore produced routines for these in Excel as the package AQB, and also routines in R, both available free on the book's website (**www.pearsoned.co.uk/barnard**), and for which we include examples of on-screen input and output for the relevant tests in the present chapter. As in previous editions of the book, we also include calculations for most of the simpler tests (Appendix II) so their mechanics are clear and they can be done by hand if required.

Test statistic

Significance tests calculate a test statistic that is usually denoted by a letter or symbol: t, H, F, χ^2, r, r_s and U are a few familiar examples from various parametric and non-parametric tests. The value of a test statistic has a known probability of occurring by chance (i.e. with random data values) for any given sample size or what are known as *degrees of freedom* (*see later*). A calculated value can thus be checked to see whether it concurs with or passes (positively or negatively, depending on the test) the threshold value appropriate to the level of probability chosen for significance. This used to mean comparing the value with a table of threshold values, but such comparisons are now made automatically for the tests in many statistical computer packages.

3.3.3 Tests for a difference

Tests of differences between groups involve asking whether there are differences in overall counts, or in mean (parametric tests) or median (non-parametric tests) values. Since non-parametric tests replace the data with their ranked values, these tests actually test for differences in the mean ranks, which, when decoded back to the original data, represent the medians. Apart from tests of counts, all these tests involve comparing the variation in values within groups with the differences in the central values (mean or median) between groups: they are therefore all forms of '*analysis of variance*'. We shall introduce three types of difference test, with parametric and non-parametric equivalents as appropriate: first χ^2 (chi-squared, pronounced 'ky-squared'), then the *t*-test and Mann–Whitney *U*-test as parametric and non-parametric (respectively) tests for differences between *two* groups, and finally other analyses of variance (parametric and non-parametric) dealing with differences between two or more groups.

Tests for a difference between two groups

We shall start with the relatively simple situation of comparing two groups. Here, two mutually exclusive groups (e.g. male/female, small/large, with property *a*/without property *a*, etc.) have been identified and measurements made with respect to each (e.g. the body length of males versus the body length of females, the number of seeds set by small plants versus the number set by large plants, the survival rate of mice on drug A versus the survival rate on drug B). Depending on the kind of measurement made, we can use one of a number of tests to see whether any differences are significant. Notice that two things are measured on each individual replicate: each can be a nominal (which group it belongs to), ordinal or constant interval variable.

Chi-squared (χ^2). A chi-squared test can be used if data are in the form of counts, i.e. if two groups have been identified and observations classified and then the total number of observations in each group counted up. Chi-squared can be used *only* on raw counts; it cannot be used on measurements (e.g. length, time, weight, volume) or proportions, percentages, or any other derived values. The test works by comparing observed counts with those expected by chance or on some prior basis. As an example, we can consider a simple experiment in Mendelian inheritance. Suppose we crossed two pea plants that are heterozygous for yellow and green seed colour, with yellow being dominant. Our expectation from the principles of simple Mendelian inheritance, of course, is that the progeny will exhibit a seed colour ratio of 3 yellow : 1 green. We can use the chi-squared test to see whether our observed numbers of yellow and green seeds differ from those expected on a 3 : 1 ratio. The expected numbers are simply the total observed number of progeny divided into a 3 : 1 ratio. Thus:

	Seed colour		
	Yellow	Green	Total
Number observed	130	46	176
Number expected	132	44	176

To find our χ^2 test statistic using R and AQB, we can follow the procedure in Box 3.2.

Box 3.2   A test comparing counts classified into two groups (1×2 χ^2 test)

In our example of Mendelian ratios in seed colour (*see text*), the data are nominal, in that each seed is classified into one of two colour groups (yellow or green). The nominal factor *colour* forms the groups, with two levels, *yellow* and *green*. We then arrive at the total numbers for the groups (frequencies of each colour) and analyse them using χ^2. In this case, the null hypothesis H_0 is that the totals will be no different from those expected under a Mendelian ratio of 3 : 1.

The data will be in one of two formats: either raw data in the form of a list of the colours of each individual seed, or the total numbers of seeds calculated for each colour (as in the table in the text). AQB deals with the ready-calculated totals.

Calculating a 1×2 χ^2 using AQB

In AQB (Fig. (i)), click on the tab at the bottom of the screen labelled 'chi-sq' and use the upper input table marked '*1-way*' (since your data are categories [*yellow/green*] of a single classification [*colour*]). Enter the calculated totals in the *data: counts* line. Since you expect a 3 : 1 ratio, divide your total number into this ratio ($\frac{3}{4}$N and $\frac{1}{4}$N) to generate the expected numbers, and enter these in the *expected* line. The result appears automatically in the 'RESULTS' table on the right. In this case, the value of χ^2 with one degree of freedom (the number of groups minus 1) is 0.12. The probability (p) associated with a value

as low as this is $p = 0.73$, which is much greater than 0.05, so we can conclude that the ratio of yellow : green seeds does not differ from our expected Mendelian ratio and we do not reject the null hypothesis.

Calculating a 1×2 χ^2 using R

If we have the data as a vector called `colour`, a sequence of individuals each scored for the group (yellow or green) to which it belongs, then the totals have to be calculated first, before applying them to the test under the 3 : 1 expected ratio. We type:

```
> z <- table(colour)
```

to get the two totals, and then (using `p=c(..)` to provide the expected ratios) we type:

```
> chisq.test(z, p=c(0.75,0.25))
```

which gives:

```
Chi-squared test for given
probabilities
data: c(130, 46)
X-squared = 0.1212, df = 1, p-value
= 0.7277
```

If we know the totals already, then we can type:

```
> chisq.test(c(130,46),
  p=c(0.75,0.25))
```

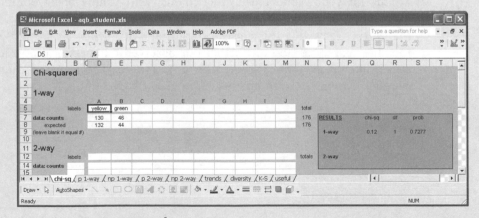

Figure (i) Performing a 1×2 χ^2 test in AQB.

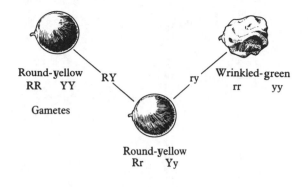

Round-yellow
RR   YY   RY   ry   Wrinkled-green
                              rr   yy
Gametes

Round-yellow
Rr   Yy

In the pea example, the expected numbers were dictated by the Mendelian theory of inheritance; we have good reason to expect a 3 : 1 ratio of yellow : green seeds and thus a difference between groups. In many cases, of course, we should have no particular reason for expecting a difference, and our expected numbers for the two groups would be the same (half the total number of observations each). Thus if our yellow and green groups had referred to pecks by chicks at one of two different coloured grains of rice on a standard background instead of the inheritance of seed colour, our chi-squared table would have looked very different:

	Grain colour	
	Yellow	Green
Number of pecks observed	130	46
Number of pecks expected	88	88

and the result would have been a χ^2 value of 40.09, which exceeds even the threshold value (10.83) for a significance level of 0.1 per cent (0.001) (*see* Appendix III, Table A). In this case we could safely reject the null hypothesis of no difference in the number of pecks to different coloured grains and infer a bias towards yellow on the part of the chicks.

An important point to bear in mind with χ^2 is that it is not very reliable when very small samples are being tested. While a correction can be brought into play here, it is a good idea not to use χ^2 when the sample size is smaller than 20 data values or when any expected value is less than 5.

t-**tests and Mann–Whitney U-tests.** *t*-tests and Mann–Whitney *U*-tests can both be performed on raw counts like chi-squared, except that they deal with each contributing data value in the two groups separately instead of as a single total. Thus if the pecking data in our last example of chi-squared were derived from ten chicks given the opportunity to peck at yellow grains and ten more given the opportunity to peck at green grains on a standard background, the values that would be used in the chi-squared and the *t*- or *U*-tests can be indicated as follows:

Chick	Pecks to yellow	Chick	Pecks to green	
a	12	k	2	
b	14	l	3	
c	13	m	10	
d	3	n	6	
e	23	o	4	A *t*- or *U*-test uses these values
f	13	p	5	
g	11	q	3	
h	15	r	1	
i	9	s	7	
j	17	t	5	
Total	130		46	A chi-squared test uses these values (or, in principle, any subtotal of data values)

In addition, however, *t*- and *U*-tests can deal with data other than counts. We can use them to compare two groups for constant-interval data such as body size, time spent performing a particular behaviour, percentage of patients responding to different drug treatments, or a host of other kinds of data. The non-parametric *U*-test can also cope with ordinal (rank) data. In addition, unlike some kinds of two-group test working on individual data values, the *U*-test and most forms of the *t*-test do not require equal sample sizes in the two groups. However, and we stress this again, these tests can be used *only* for comparing *two* groups and only if each data value is *independent* of all other data values. Furthermore, the *U*-test can only test a general (or two-tailed) prediction (i.e. there is a difference, but not in any one predicted direction). Analyses for differences between two groups using AQB and R are shown in Boxes 3.3a–d.

Tests for two related samples

The tests for two groups introduced so far assume that data in the different groups are independent of each other. In many cases, the values of the data in each group are not independent, but are related in some way. For example, we might have tested one leaf of a plant with one treatment, and another leaf of the same plant with a different treatment. Or we might have provided a sample of ten female Siamese fighting fish (*Betta splendens*) with a red male as a potential mate on one occasion, and a blue male on another occasion, and measured their reactions to each of them. In order to test for the effect of different treatments with these kinds of data, we need a method that takes the *non-independence* of the data into account. In these kinds of experiments, the data usually come in replicated pairs (as in the 'pair' of males presented to each female above), and the appropriate statistical tests are known as 'paired-sample' tests. The sample sizes of the two groups are therefore necessarily the same.

The methods that deal with this design work by taking the differences between the measurements of each pair, and then analysing these differences. In the parametric case, the *paired t-test* tests whether the mean of these differences is significantly different from zero. In the non-parametric case, the *Wilcoxon matched-pairs signed ranks test* ranks all the differences, ignoring whether they

Box 3.3a Mean values: a *general* parametric test for two groups (two-tailed *t*-test)

We shall use data for chicks pecking at yellow and green grains of rice. The data being analysed are thus measured on a constant interval scale – the number of pecks. Testing the residuals for normality (*see* Box 3.1b) shows that they do not depart significantly from a normal distribution (Shapiro–Wilk = 0.98, d.f. = 80, ns), so we can analyse them perfectly reasonably using a parametric test. The nominal factor *seed colour* constitutes the grouping, with two categories, *yellow* and *green*.

First we frame our prediction, in this case the general prediction H_1 that chicks will peck at different rates towards the two colours of grain; i.e. that the two groups

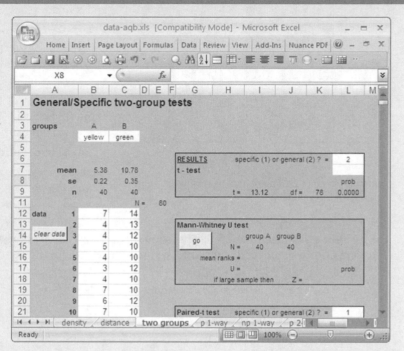

Figure (i) Data entry and analysis output for comparing the means of two groups in a two-tailed test in AQB.

yellow (A) and *green* (B) will have different mean values (A ≠ B). Thus the null hypothesis H_0 is that the two mean values will not differ.

Testing for a general (two-tailed) difference in the means of two groups using AQB

In AQB (Fig. (i)), select the 'two groups' worksheet from the options at the bottom of the screen, and enter a '2' (for a general test) in the top portion of the top right-hand square in the first ('RESULTS, *t*-test') box. Then enter the data for each group in a separate column. The mean values, standard errors and sample sizes will automatically appear, together with the test statistic '*t*' in the box labelled '*t*-test'. This value for *t* allows for the possibility that each group may have a different variance. The value of *t* is high (13.12), with a *p*-value of less than 0.0001, so we can reject the null hypothesis of no difference between the two groups.

Testing for a general (two-tailed) difference in the means of two groups using R

In R, we enter the data in the standard format, with the data values in a single column (`pecks`), and the group to which each value belongs coded in another column (`colour`). First plot the data (Fig. (ii)), using the routines we wrote for standard errors and plotting (see p. 40):

```
> labels <- as.character(levels(colour))
> mpecks <- tapply(pecks, colour, mean)
> sepecks <- tapply(pecks, colour, se)
> plot.error.bars(mpecks, sepecks, labels)
```

The difference seems very large. The function in R that tests this difference automatically assumes that the variances of the two groups are not equal (a Welch *t*-test), and so we test whether that is true, first

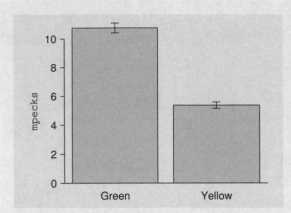

Figure (ii) Mean (± s.e.) number of pecks made by chicks to green and to yellow rice.

to tell us whether we should suppress this assumption or not. We have a look at the variances first:

```
> tapply(pecks, colour, var)
    green    yellow
4.794231 1.983974
```

The variance in pecks for the green group is more than twice that of the yellow group. We use the Bartlett test again (see Box 3.1b) to see whether this is a significant difference in variances:

```
> bartlett.test(pecks ~ colour)
Bartlett test of homogeneity of variances
data: pecks by colour
Bartlett's K-squared = 7.2629,
df = 1, p-value = 0.00704
```

The *p*-value is significant, indicating we should reject the null hypothesis of equal variances; the variances are clearly significantly different. We therefore either specifiy, or do not alter, the default option (var.equal=FALSE) for the *t*-test:

```
> t.test(pecks ~ colour, var.equal = FALSE)
Welch Two Sample t-test
data: pecks by colour
t = 13.118, df = 66.559, p-value < 2.2e-16
alternative hypothesis: true difference in means is not equal to 0
95 percent confidence interval:
 4.578244 6.221756
sample estimates:
 mean in group green mean in group yellow
            10.775                5.375
```

Notice that the value of *t* is the same, but the value for the degrees of freedom is no longer an integer, but is adjusted for the unequal variances. This affects the actual value of *p*, but here does not affect our conclusion.

Box 3.3b   **Mean values: a *specific* parametric test for two groups (one-tailed *t*-test)**

We will use here the same data as in Box 3.3a, chicks pecking at yellow and green grains. Once more, we frame our prediction; this time, though, we are going to predict that one particular group will have a larger mean than the other – i.e. we shall decide *in advance of collecting the data* which of the two groups (A and B) is predicted to have the greater mean value. There needs to be

some a priori reason (theory, or previously published or gathered data) for this prediction; it can't just be made on a whim. Suppose that, on the basis of your knowledge, you predict that chicks should peck more at yellow than at green seeds, i.e. that A > B (H_1). The null hypothesis (H_0) is that the mean value of A is not greater than that of B.

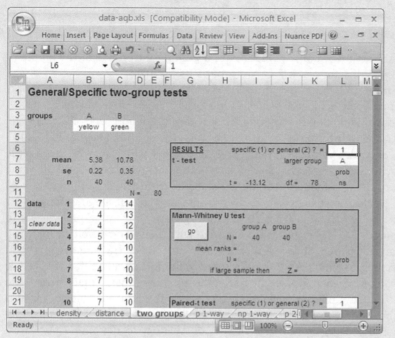

Figure (i) Data entry and analysis output for comparing the means of two groups in a one-tailed test in AQB.

Testing for a specific (one-tailed) difference in the means of two groups using AQB

In the 'two groups' worksheet of AQB again (Fig. (i)), choose a specific test by inserting '1' in the top half of the box in the top right corner of the first 'RESULTS' box, and enter the group expected to have the larger mean value (in this case, 'A') in the lower half. Notice that the value of t is large (-13.12) but negative (Fig. (i)), indicating that, contrary to prediction, the mean of B is larger than that of A. Because of this, the p-value says 'ns', i.e. not significant. Had you predicted the order the other way round (B > A), then you would have obtained a significant result (if you replace 'A' by 'B' in the lower half of the test selector box, you will see that the t-value becomes positive, and the probability changes from 'ns' to 0.0000, actually half the two-tailed probability), but as it is, the result is not significant, *and you most certainly cannot change the prediction retrospectively to make it significant.*

Testing for a specific (one-tailed) difference in the means of two groups using R

In R, we can only do a one-tailed test if the data for the two groups are in separate columns. So we will learn how to extract them from our single column of data (`pecks`) using the group names (`colour`). We first extract into the vector 'yellow' all the values of pecks for which `colour = yellow` is `TRUE`, and then the same for green. Notice the square brackets, indicating the rows of the vector:

```
> yellow <- pecks
  [colour=="yellow"]
> green <- pecks
  [colour=="green"]
```

Now we have them in two separate vectors, and can do the one-tailed t-test. The vectors you provide x, y are followed by what the hypothesis is: `alternative = c("greater")` means that the alternative to the null hypothesis is that $x > y$. Other possibilities are `alternative = c("less")`, and `alternative= c("two.sided")` – the default. Our hypothesis is that `yellow > green`:

```
> t.test(yellow, green,
  alternative=c("greater"))
```

which gives:

```
Welch Two Sample t-test
data: yellow and green
t = -13.118, df = 66.559, p-value = 1
alternative hypothesis:
true difference in means is greater
than 0
95 percent confidence interval:
 -6.08666 Inf
sample estimates:
mean of x mean of y
   5.375    10.775
```

Box 3.3c Mean ranks (medians): a *general* non-parametric test for two groups (Mann–Whitney *U*, Wilcoxon rank sum test)

This test is sometimes known as the Mann–Whitney, and sometimes as the Wilcoxon rank sum test. The calculations differ for each, but it is essentially the same test. Suppose an ecologist was interested in the effect of microhabitat on the distribution of periwinkles (*Littorina* spp.) on a rocky shore. Two habitats – a boulder/shingle beach and crevices in a rocky stack – were compared for the prevalence of the commonest periwinkle species measured as the percentage of the total number of all invertebrate individuals recorded within quadrat samples.

The data being analysed are measured on a constant-interval scale – percentage – but these are bounded by zero and 100, and are therefore unlikely to be normally distributed unless all values are well away from the boundaries. Testing the residuals for normality (as in Box 3.1b) shows that the distribution is indeed far from normal (Shapiro–Wilk = 0.94, d.f. = 80, $p < 0.001$), which remains so even when the data have been arcsine-square-root transformed (*see* p. 61) to 'stretch out' the tails: hence a non-parametric test is required. The nominal factor *habitat* forms the level of grouping, with two categories: shingle *beach* and *rock* crevices.

Frame the prediction, in this case the general prediction H_1 that the prevalence of periwinkles will differ between the two kinds of habitat. In terms of our non-parametric test, we are predicting that the two groups *beach* (A) and *rock* (B) will have different mean ranks (A ≠ B). Thus the null hypothesis H_0 is that the two mean ranks will not differ.

Testing for a general difference in the medians of two groups using AQB

In AQB (Fig. (i)), choose the 'two groups' worksheet as in Boxes 3.3a, b, enter the data for each group in a separate column, and click the 'go' button of the lower box labelled 'Mann–Whitney *U*-test'. This will give you the sample size in each group, the mean ranks, and the value of the test statistic, *U* (for small sample sizes < 20 in both groups) and *z* (for large sample sizes > 20 in either group). The value of *U* needs to be compared with the critical values for *U* in Appendix III, Table B. If the value is less than the threshold value for a probability of 0.05, we can reject the null hypothesis that there is no difference between the groups. Note that in this test we use the sample sizes N_A and N_B rather than the degrees of freedom to determine our threshold value. In this case, we have large sample sizes, and AQB has calculated *z*, and provided

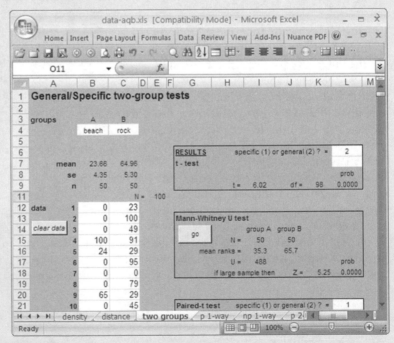

Figure (i) Data entry and an analysis output for comparing the medians of two groups in a general test in AQB.

automatically the probability under the null hypothesis. Since $z = 5.25$ ($p < 0.001$), the difference in mean ranks is obviously highly significant. We ignore the sign of the z-value here because we have posed a general rather than a specific prediction, and the sign is only relevant to a test of a specific prediction.

Testing for a general difference in the medians of two groups using R

In R, the data (`periwinkle`) are in a single column, and the group to which each value belongs is in another column (`habitat`). There are two names for this test, the Mann–Whitney and Wilcoxon rank sum test, and R uses the latter name. We first plot the data (Fig. (ii)), checking first to see that habitat is recognised by R as a factor:

```
> is.factor(habitat)
[1] TRUE
> plot(habitat, periwinkle)
```

The medians seem very different, which is encouraging. Then we invoke the test:

```
> wilcox.test(periwinkle ~ habitat)
```

which gives:

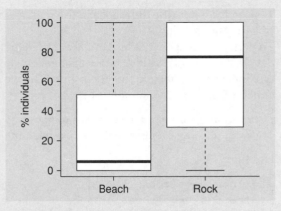

Figure (ii) Boxplot of the medians (with interquartile ranges and extremes) of the prevalence (% of number of individuals in a sample) of periwinkles (*Littorina* spp.) in two shore habitats (shingle beach or rocky stack).

```
Wilcoxon rank sum test with
continuity correction
data: periwinkle by habitat
W = 488, p-value = 1.047e-07
alternative hypothesis:
true location shift is not equal to 0
```

Since the p-value is much less than 0.001, then the difference in mean ranks is highly significant.

Box 3.3d Mean ranks (medians): a *specific* non-parametric test for two groups

We shall analyse the same data as in Box 3.3c for this test. Assume that from the literature the ecologist is aware that periwinkles are said to form a greater proportion of the invertebrate community in rocky habitats. The prediction is therefore that periwinkles will show a higher prevalence in the *rock* habitats than in the *beach* habitats, i.e. H_1 is that B > A. Thus the null hypothesis is that B is not greater than A, i.e. H_0 is that B ≤ A.

Testing for a specific difference in the medians of two groups using AQB

In AQB, unless there are large enough sample sizes (n for either group > 20) so that the z test statistic is calculated, it is not possible to do a specific non-parametric version of the Mann–Whitney. If a z-score is available, and the difference is in the predicted direction, then merely halve the probability to get the specific (one-tailed) probability. In the data set here, notice that the mean rank for B (65.7) is indeed greater than the mean rank for A (35.3). It is not uncommon for the order of the arithmetic means to be round the other way to the mean ranks, but here they should be the same. The mean ranks are in the predicted pattern, and hence we can halve the probability of the z-score (which is available because of the large samples). Again, the result is highly significant.

Testing for a specific difference in the medians of two groups using R

To perform a specific test in R, we again need the data in two separate vectors rather than in a single one indexed by a factor. Thus as in Box 3.3b, we convert from one column into two new vectors:

```
> rock <- periwinkle[habitat=="rock"]
> beach <- periwinkle[habitat=="beach"]
```

and then

```
> wilcox.test(rock, beach,
  alternative=c("greater"))
```

Wilcoxon rank sum test with continuity correction
data: rock and beach
W = 2012, p-value = 5.237e-08
alternative hypothesis:
true location shift is greater than 0

If the prediction had been the other way (rock < beach), then:

```
> wilcox.test(rock, beach,
  alternative=c("less"))
```

Wilcoxon rank sum test with continuity correction
data: rock and beach
W = 2012, p-value = 1
alternative hypothesis:
true location shift is less than 0

Definitely not significant!

are positive or negative, and then tests whether the mean rank of the positive differences is the same as that of the negative ones. A further test is also possible here, since the null hypothesis is that there are equal numbers of positive and negative differences: the *binomial test* tests for a departure from this equality. Boxes 3.4a–d show how to do some of these tests in AQB and R.

In fact, having data values in pairs like this is only a particular case of data occurring as *related samples*, where data values are not independent of each other for a wide range of reasons, for instance because subjects have been kept in the same cage and have been exposed to the same social experience that is different from the social experience in any other cage. Such sources of non-independence lead to data being treated in so-called *blocks*, where the blocking designation allows the non-independence to be taken into account in subsequent analyses. The experimenter may be completely uninterested in whether or not there is a significant block effect, but it must be allowed for in the analysis in order to see the true impact of the treatment. Box 3.5e shows an example of a more general related samples (or 'repeated measures') analysis.

Tests for a difference between two or more groups

So far, we have introduced significance tests that can test for a difference between two groups. In many cases, of course, we shall be faced with more than two groups. For instance, Fig. 2.4 suggests that the amount of damage around the edges of leaves increases with leaf size class, perhaps indicating a predilection for big leaves by particular kinds of pest. If we wanted to know whether the difference between the size classes was significant, we should have to deal with three groups of data. How do we do it? The temptation to which many succumb is to do a *round robin* comparison of pairs of groups using a *t*- or *U*-test. In our Fig. 2.4 example, this would mean testing for a difference between small and medium leaves, then for a difference between medium and large leaves, and finally for a difference between small and large leaves. *The error of this cannot be emphasised too strongly*.

Box 3.4a Mean values: a *general* parametric test for *related* samples in two groups

A botanist wanted to assess whether leaves treated with nitrogen differed in length from untreated leaves. She therefore treated one leaf with water and another leaf of the same plant with a solution of nitrogen, replicating this across 40 plants. After a suitable length of time, she measured the length of each leaf.

The data being analysed are measured on a constant-interval scale – length. The nominal factor 'treatment' forms the groups, with two categories: water-treated (*control*) and nitrogen-treated (*N*). The (nominal) blocking factor (*see text*) that relates values in the two groups is plant individual, with 40 levels. A related-samples test looks at the differences between each pair of values, and it is these differences that should be normally distributed for a parametric test. The differences need to be calculated in R into a new variable, and then this tested in the usual manner using `shapiro.test(differences)`. When we do this, we find that the differences can indeed be assumed to be normally distributed (Shapiro–Wilk = 0.98, $n = 40$, ns).

Frame the prediction. Here it is a general prediction that the treatment will change the length of the leaf within plants, so that water-treated controls (A) and nitrogen-treated (B) leaves will differ in length, i.e. H_1 is that A ≠ B. H_0 is therefore that A = B.

Testing for a general difference between related samples in two groups using a parametric test (two-tailed paired *t*-test) in AQB

In the 'two groups' worksheet of AQB, choose the output table labelled 'Paired *t*-test' and enter 2 (for a general test) in the top right corner box, then enter the data for each group in separate columns. Since the data are paired, make sure that the data for each pair (in this case, each individual) are in the same row, and therefore that there is exactly the same number of data values in each column. The paired *t*-test result will appear automatically in the output box (Fig. (i)). The difference between the paired values amounts to an average of 4.65 ± 0.42 mm ($n = 40$ pairs), which is significantly different from zero ($t = 11.2$, d.f. = 39, $p < 0.001$). The conclusion, therefore, is that treating leaves with nitrogen significantly changes the length of leaves.

Figure (i) Data entry and output for a two-tailed paired *t*-test in AQB.

Testing for a general difference between related samples in two groups using a parametric test (two-tailed paired *t*-test) using R

For paired samples we need the data in two separate vectors because R will pair them by row. The rows are called 'control and 'nitrogen'. First we obtain the means and standard errors of the two groups using the routine for standard errors we wrote before (p. 40), and then plot them (Fig. (ii)) – note these values take no account of the pairing of the data:

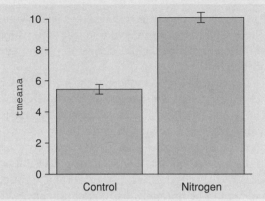

```
> m1 <- mean(control)
> m2 <- mean(nitrogen)
> tmeans <- c(m1, m2)
> se1 <- se(control)
> se2 <- se(nitrogen)
> ses <- c(se1, se2)
> labels <- c("control", "nitrogen")
> plot.error.bars(tmeans, ses, labels)
```

Figure (ii) Mean (± s.e.) leaf length (mm) for control and nitrogen-treated leaves.

If we wanted to have a look at the impact of the pairing, then we would take the differences between every pair, and plot the mean and SE of the differences, again using the plot routine we wrote before (p. 40).

```
> diffs <- nitrogen - control
> plot.error.bars(mean(diffs), se(diffs), c("differences"))
```

The result (Fig. (iii)) is encouraging because the difference looks large compared to its SE. Then to run the test we type:

```
> t.test(control,nitrogen,
  paired=TRUE)
```

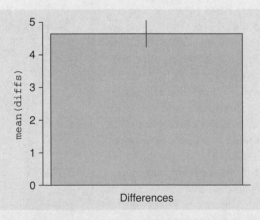

```
Paired t-test
data: control and nitrogen
t = -11.1139, df = 39, p-value =
1.183e-13
alternative hypothesis:
true difference in means is not equal
to 0
95 percent confidence interval:
 -5.490376 -3.799624
sample estimates:
mean of the differences
               -4.645
```

Figure (iii) Mean (± s.e.) difference in length (mm) between paired control and nitrogen-treated leaves.

Thus there is a highly significant effect. The value is negative because the 'nitrogen' values are larger. Had we put the vectors the other way round, then *t* would have been positive.

Box 3.4b Mean values: a *specific* parametric test for *related* samples in two groups

Using the same example of leaf length as Box 3.4a, decide *in advance of collecting the data* which of the two groups (A or B) is predicted to have the greater mean value. As always, there needs to be some a priori reason (theory, or previous published or gathered data) for this prediction. Suppose that, on the basis of her knowledge, the botanist in Box 3.4a predicted that a leaf treated with nitrogen would be *longer* than the control leaf on the same plant, i.e. that B > A (H_1). The null hypothesis (H_0) is that the mean value of B would not be greater than that of A.

Testing for a specific difference between related samples in two groups using a parametric test (one-tailed paired *t*-test) in AQB

In AQB, using the 'two groups' worksheet, choose a specific test by inserting '1' in the top right corner of the 'Paired *t*-test' output box (see Box 3.4a, Fig. (i)), and select which group is expected to have the larger mean value ('B'). Notice that the mean difference changes from negative to positive, reflecting the fact that the differences are now taken to be B – A rather than the default A – B. The value of *t* also changes from negative to positive, and the significance changes from 'ns' (caused by the combination of a specific test but a negative *t*-value) to '0.0000' (or < 0.0001, as we should cite it). The difference is clearly highly significant in the predicted direction.

Testing for a specific difference between related samples in two groups using a parametric test (one-tailed paired *t*-test) in R

This is very simple to do in R. It is almost exactly the same as doing a two-tailed test, with an extra argument:

```
> t.test(control,nitrogen,
    paired=TRUE, alternative=c("less"))

Paired t-test
data: control and nitrogen
t = -11.1139, df = 39, p-value =
5.916e-14
alternative hypothesis:
true difference in means is less
than 0
95 percent confidence interval:
    -Inf -3.940813
sample estimates:
mean of the differences
              -4.645
```

The output is very similar except the *p*-value is even lower than the two-tailed version (because a one-tailed test is more powerful).

The conclusion from both tests is that the data support the idea that nitrogen-treated leaves grow to be longer than control leaves on the same plant (paired $t = 11.1$, d.f. = 39, $p < 0.001$), and the null hypothesis can be rejected.

The most serious problem arising from such a practice is that it increases the likelihood of obtaining a significant difference by chance when really none exists. To take an extreme example: if we carried out 100 two-group comparisons, then, just by chance, five of them stand to be significant at the 5 per cent level. Even if we made only 20 comparisons, one is likely to be significant by chance. While this may not seem a serious difficulty when we are dealing with only three or four groups, these examples illustrate the error in principle. To get round the problem, we need tests that can cope with comparisons between several groups at the same time.

One-way analysis of variance. Where we have series of data values falling into several groups (in a similar fashion to data values in the two groups of a *t*- or

Box 3.4c   Mean ranks: a *general* non-parametric test for related samples in two groups (Wilcoxon matched-pairs signed ranks test)

In an experiment on two mouse strains, the average number of offspring was compared over 50 years. The data being analysed are measured on a constant interval scale – the average number of offspring. The nominal factor *strain* forms the level of grouping, with two levels: *strainA* and *strainB*. The (nominal) blocking factor that relates values in the two groups is year, with 50 levels. As mentioned before (Box 3.4b), the normality assumption in related-samples tests concerns the differences between the related values. When the differences are tested here, there is strong evidence of a non-normal distribution (Shapiro–Wilk = 0.88, $n = 50$, $p < 0.001$). Thus a non-parametric test is appropriate.

Frame the prediction. The general prediction is that there will be a difference in the number of offspring between the two strains across years. Thus, allowing for the related samples (i.e. the differences among years), the prediction (H_1) is that A ≠ B.

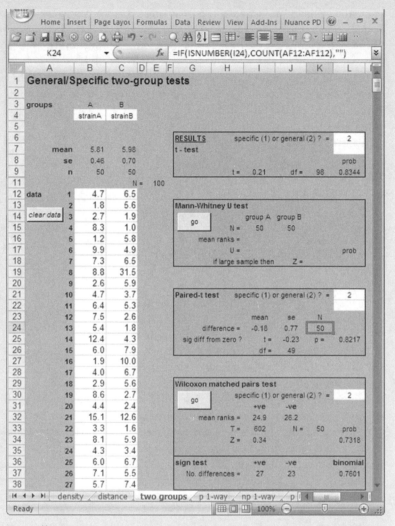

Figure (i) Data entry and output for a general Wilcoxon matched-pairs signed ranks test in AQB.

Testing for a general difference between related samples in two groups using a non-parametric test in AQB

In the data in the 'two groups' worksheet in AQB (Fig. (i)), each year is a row in the table. Clicking 'go' in the 'Wilcoxon matched-pairs test' box shows that the mean of the positive ranks is 24.9, and the mean of the negative ranks is 26.2. The test statistic is T, and it has a value of 602 (note that T is different from t of the t-test). However, since the sample size is large enough, a z-value is calculated and this is the value we use. It is not large enough (0.34) to reject the null hypothesis, and thus the result is not significant.

Testing for a general difference between related samples in two groups using a non-parametric test in R

As before, the data are in two separate columns ('strainA' and 'strainB') so that they can be paired by row, with each row representing a block (year). We first plot the median of the differences (Fig. (ii)), to see whether it is plausible that they differ:

```
> diff <- strainA - strainB
> fac <- factor(rep(c("diff"),
  times=length(diff)))
> plot(fac,diff,ylab=c("difference"))
```

The median is close to zero, and the range clearly encloses zero: this is not promising. Then invoke the test:

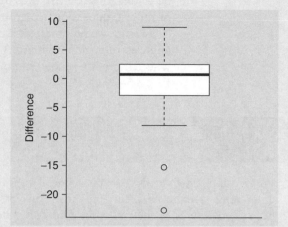

Figure (ii) Median (with interquartile ranges and extremes) difference in the number of offspring between mouse strains in any one year.

```
> wilcox.test(strainA, strainB, paired=TRUE)
```

```
Wilcoxon signed rank test with continuity correction
data: strainA and strainB
V = 675.5, p-value = 0.7173
alternative hypothesis: true location shift is not equal to 0
```

The calculation is slightly different from that in AQB, but the result is the same. There is no evidence here for a difference between these two strains.

Box 3.4d Mean ranks (medians): a *specific* non-parametric test for related samples in two groups

The specific version of the AQB test merely halves the probability, and it is up to you to check that the mean ranks are in the predicted direction. In R, you simply modify the command, as before. Here we test whether strainA > strainB:

```
> wilcox.test(strainA, strainB, paired=TRUE,
  alternative=c("larger"))
```

U-test), a one-way analysis of variance (often expressed as the acronym *one-way ANOVA*) is a suitable significance test. There are both parametric analyses of variance and non-parametric tests that have an equivalent function. We shall introduce both kinds here. While parametric analyses of variance are by far the more widely used, the non-parametric version we shall describe has the advantage not only of being robust to a wider range of data but also of allowing two kinds of specific predictions about the *direction* of differences between groups to be tested. This test can thus be used to test general or specific predictions for two or more groups. For two groups it is therefore more flexible than the *U*-test and, as a result, we recommend it even in these cases. Procedures for the two kinds of analysis of variance in AQB and R are outlined in Boxes 3.5a–e. Procedures where there is a related samples design with more than two groups are shown in Box 3.5e.

Box 3.5a   Mean values: a *general* parametric one-way ANOVA for differences between two or more groups

We are interested in developing artificial techniques for rearing bumblebees (*Bombus* spp.) for pollinating greenhouse plants. Suppose we measure the weights of bumblebee queens that have been overwintered in the soil under four different conditions. The first condition is the normal rearing environment (soil), while different components (stones, leaves or cotton wool) are added to the rearing environment of the other groups. We think that any of these additional components might increase the weight of the resulting queens, and heavier queens in spring have a better chance of producing successful colonies.

The data being analysed are the weights of queens. The nominal factor *treatment* forms the level of grouping, with four categories (1 = normal soil, 2 = plus stones, 3 = plus leaves, 4 = plus cotton wool). A test of the residuals (see Box 3.1b) shows that we can assume that they are normally distributed (Shapiro–Wilk = 0.996, d.f. = 160, ns).

Frame the prediction, in this case the general one that there are differences among the mean values of the treatment groups (H_1). The null hypothesis (H_0) is therefore that there are no differences among the groups.

Testing for a general difference in the means of two or more groups using a parametric one-way ANOVA in AQB

In AQB the data are entered as four columns, the data for each group in a separate column (Fig. (i)). The mean values and their standard errors appear above each column, and can be pasted into Excel for plotting (see Box 2.3). The results appear automatically in the 'one-way ANOVA table (general)' in the RESULTS box to the right. It is set out in the conventional format of an analysis of variance output table, which it is standard practice to reproduce in reports and papers as the result of the test. The test statistic is *F*, which reflects the ratio of between-group to within-group variation (hence it is usually referred to as an *F*-ratio): since each of these sources of variation has a degree of freedom value, the test statistic has two sets of degrees of freedom. Thus, in this case, we should report *F* as $F_{3,156} = 3.31$, $p < 0.05$ (or $F = 4.31$, d.f. = 3,156, $p < 0.05$).

Testing for a general difference in the means of two or more groups using a parametric one-way ANOVA in R

The data should be in the usual format for independent samples, i.e. the data values in one column, and group codes in another. If the data for each group are in a separate column, don't despair, because you can join them all together like this:

```
> bbee <- c(control, stones, leaves, cotton)
```

and then generate a factor with the group codes from the names of the variables:

```
> treatment <- as.factor(rep(c(names(dtfr)),each=40))
```

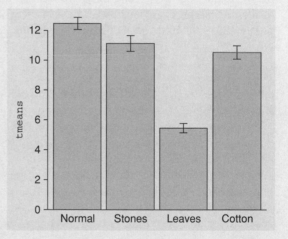

Figure (i) Data entry and results output for a general one-way ANOVA in AQB.

We first plot the data, using the routines we wrote (Box 2.3), and getting the labels from the variable names in the dataframe:

```
> tmeans <- tapply(bbee, treatment,
  mean)
> ses <- tapply(bbee, treatment, se)
> labels <- names(dtfr)
> plot.error.bars(tmeans, ses,
  labels)
```

The plot (Fig. (ii)) seems to show large differences among groups, especially for the 'leaves' treatment.

Now we have everything we need, but we should check for equal variances first:

```
> fligner.test(bbee~treatment)
```

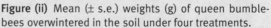

Figure (ii) Mean (± s.e.) weights (g) of queen bumble-bees overwintered in the soil under four treatments.

```
Fligner-Killeen test of homogeneity
of variances
data: bbee by treatment
Fligner-Killeen:med chi-squared = 4.0447, df = 3, p-value = 0.2567
```

This is not significant, and hence there is no evidence to reject the assumption of homogeneity of variances: we already know that the residuals are normal. The ANOVA is then very straightforward. The command 'aov' does the ANOVA, with a response variable (bbee) being predicted (the '~'

symbol) by a categorical predictor, i.e. a factor (treatment). The command 'summary' puts it in a convenient format:

```
> summary(aov(bbee ~ treatment))

            Df  Sum Sq  Mean Sq  F value  Pr(>F)
treatment    3   80.28  26.7586   3.3131  0.02163 *
Residuals  156 1259.95   8.0766
---
Signif. codes: 0 '***' 0.001 '**' 0.01 '*' 0.05 '.' 0.1 ' ' 1
```

We get the ANOVA table in standard format: when reporting an ANOVA, it is often a good idea to reproduce this table because it says everything you need to know.

The conclusion in both AQB and R is therefore that there is good evidence that the treatments affect the weight of overwintering queen bumblebees ($F = 3.31$, d.f. = 3,156, $p < 0.05$).

Post-hoc multiple comparisons

A general one-way ANOVA will tell us whether or not there is a significant difference between our groups, but it does not tell us where the difference lies. We cannot conclude which pairs of mean values are significantly different (although we can certainly suggest them from the group means), because we did not make any specific prediction beforehand, in advance of obtaining the data, about the ordering of the mean values, nor did we set up planned contrasts between particular sets of groups. Of course, it is entirely natural to want to delve further and find out where the differences lie, and many researchers would like to be able to make such *post hoc* tests. A plethora of different tests is available that test every mean value against every other one. Crawley (2007: 483–485) recommends the Tukey Honest Significant Difference, implemented in R by typing

```
> TukeyHSD(aov(bbee - treatment))
```

The very diversity of methods ought to warn us that any one of them might be inadequate. Many biologists use them routinely, but others, and many statisticians, avoid them altogether. Detailed study shows that none of them can be relied upon to give accurate *p*-values for differences (Day & Quinn, 1989): many are too conservative, but others are not conservative enough, suggesting significance unjustifiably. Thus our view is that, if they are used at all, they should be taken only as a rough guide to where the differences might lie (so are not really much better than simply inspecting the group means). None can be recommended over the others. Many people use the Least Significance Difference (LSD). All this does is to take the within-groups ('error') mean square from the analysis of variance and to construct a 95 per cent confidence limit with it. However, the LSD 'test' is only accurate for a priori contrasts (Day & Quinn, 1989).

Box 3.5b Mean values: a *specific* parametric one-way ANOVA

Suppose we have measured the light absorbance of an indicator of a particular biochemical reaction under each of four temperature conditions. We were unable to arrange the experiment so as to have equal numbers of replicates (which is always desirable), but we do have several replicates for each group. The data being analysed – light absorbance – are on a constant-interval scale. The nominal factor *temperature* forms the level of grouping, with four levels (A = 10 °C, B = 15 °C, C = 20 °C and D = 25 °C). Note that although the temperature levels would be treated as nominal groups in a general ANOVA, here they not only form a rankable series, but they actually consist of constant-interval measurements that are equally spaced (by 5 degrees). This will become important in a moment. Testing the residuals for normality (Box 3.1b) shows that we can assume normality (Shapiro–Wilk = 0.97, d.f. = 35, ns), and the variances appear to be homogeneous (Fligner chi-square = 0.88, d.f. = 3, ns).

When it comes to framing our predictions, theory (and past practice) tells us that temperature speeds up biochemical reactions, and, furthermore, that a 10 °C rise in temperature should double the rate of reaction. We therefore predict that the rank order of the mean values of light absorbance should be A < B < C < D. Unfortunately, such a simple specific prediction is hard to test because it says nothing about the magnitude of the changes between the groups. It is testable using something called isotonic regression (Gaines & Rice, 1990), but this is complicated and therefore not available in AQB, nor indeed in most packages except R.

Theory says there should be a particular relationship between temperature and the rate of reaction. In this experiment with light absorbance, which is measured on a logarithmic scale, as our measure, this translates into a prediction of a *linear positive* relationship between temperature and light absorbance. Because (and only because) the temperature groups are evenly spaced, we can test this prediction using so-called polynomial coefficients (which require even spacing between the levels), using R.

Testing for a specific difference in the means of two or more groups using a parametric one-way ANOVA in AQB

In both AQB and R we can use the alternative of independent contrasts to test almost the same prediction. These make a contrast between two subsets of the groups (the factor levels). If the linear positive relationship between temperature and light absorbance is correct, then the mean light absorbance of the 10 °C group will be less than the average of the other groups. This can be tested using a contrast.

We have three degrees of freedom to use, and, since each contrast involves one degree of freedom, we can frame three independent (so-called orthogonal) contrasts. We can therefore create an hierarchical set of contrasts that encapsulates the linear relationship that actually we would like to test. We therefore predict that:

	A < (B + C + D)/3
and	B < (C + D)/2
and	C < D

We enter the data as columns in the usual manner (Fig. (i)a). However, we now need to enter some coefficients (weightings) for the first contrast in the line marked 'Contrast', and this requires a little explanation.

Basically, the contrast coefficients need to be integers that sum to zero, as indicated by the 'checks on contrasts' (cell M6) to the right of the coefficient input box. In the first contrast (for A), the prediction is:

	A < (B + C + D)/3
which is	3A < B + C + D

which, rearranged to express the inequality as '0<', becomes:

0 < –3A + B + C + D

Therefore the coefficients are –3, +1, +1 and +1 on the groups A, B, C and D, respectively. Placing these in the appropriate boxes brings up the result for the contrast ('RESULTS' box in Fig. (i)a). The ANOVA in the box shows a significant outcome ($F_{3,34} = 33.21$, $p < 0.0001$) for the general test among all the groups. Here we are not merely making a contrast, but are also predicting its *direction*, so we choose

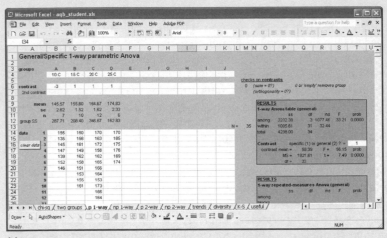

(a)

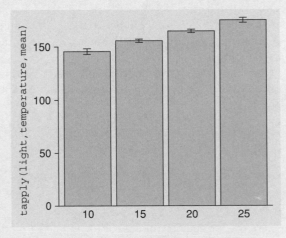

One-way ANOVA table

	ss	df	ms	F/t	prob
Among	3232.39	3	1077.46	33.21	< 0.001
A < B + C + D	1821.61	1	1821.61	t = 7.49	< 0.001
B < C + D	997.33	1	997.33	t = 5.54	< 0.001
C < D	413.44	1	413.44	t = 3.57	< 0.01
Within	1005.61	31	32.44		
Total	4238.00	34			

(b)

Figure (i) (a) Data entry and results output in AQB for a parametric one-way ANOVA using specific contrasts, (b) the resulting ANOVA and specific contrasts.

(Box 2.3). See how we can do this all in one command, rather than build the elements separately (as in Box 3.5a). The only thing to be careful about are the labels, because R orders these alphabetically, which might not be the same order as they occur in the data – it will then mismatch labels to bars. Here the levels are numbers, which order correctly:

```
> plot.error.bars(tapply(light,
    temperature,mean),
tapply(light,temperature,se),
    levels(temperature))
```

The result (Fig. (ii)) appears to show a gradually increasing mean with temperature. Then we carry out an ANOVA and ask just for a summary, to see whether there are in fact any significant differences among our groups to explore using specific predictions:

the specific version of the contrast test by entering '1' in the small white box to the lower right of the 'RESULTS' box. When we do this, a t-value and associated probability for the specific test is then generated, showing a significant fit to our directional prediction ($t_{33} = 7.49$, $p < 0.0001$).

Then we go through each of the remaining contrasts, entering the coefficients and obtaining the general ANOVA table and specific-contrast t-value (Fig. (i)b). By the same logic as above, the coefficients are 0, −2,1,1 (for 2B < C + D) and 0,0,−1,1 (for C < D).

Testing for a specific difference in the means of two or more groups using a parametric one-way ANOVA in R

The data are read in in the usual manner, as a column of the data (light) and the group codes (temperature). We declare temperature to be a factor first.

```
> temperature <- factor(temperature)
```

First we plot the data, using the routines we wrote

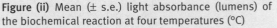

Figure (ii) Mean (± s.e.) light absorbance (lumens) of the biochemical reaction at four temperatures (°C)

```
> summary(aov(light ~ temperature))
```

and we get the same result as AQB:

```
           Df Sum Sq Mean Sq F value    Pr(>F)
temperature 3 3232.4 1077.46  33.215 8.314e-10 ***
Residuals  31 1005.6   32.44
```

So there are differences to explore among these groups.

For this example we have discussed three ways of carrying a test of a specific hypothesis on these data:

(a) using isotonic regression to test the hypothesis that A < B < C < D;
(b) using polynomial contrasts to test for a linear increase in the response variable with each level of the factor (testable only because these levels represent evenly spaced values of a constant interval measurement);
(c) using contrasts to test the three predictions that A < (B + C + D)/3, B < (C + D)/2 and C < D.

We can do all three in R.

Isotonic regression

Isotonic regression is not for the faint-hearted, and therefore we omit it here.

To use polynomial contrasts

First we have to tell R that our factor is a set of ordered levels:

```
> t2 <- ordered(temperature)
```

We can check that it really is ordered:

```
> is.ordered(temperature)
[1] TRUE
```

Then we can ask R to do the ANOVA using these ordered levels, and get the summary of the linear model to see the contrasts:

```
> summary.lm(aov(light ~ t2))

aov(formula = light ~ t2)
Coefficients:
            Estimate Std. Error t value Pr(>|t|)
(Intercept) 160.16786    0.99962 160.229   <2e-16 ***
t2.L         21.65685    2.19446   9.869 4.39e-11 ***
t2.Q          0.06905    1.99924   0.035    0.973
t2.C          0.46106    1.78277   0.259    0.798
Residual standard error: 5.696 on 31 degrees of freedom
Multiple R-squared: 0.7627,     Adjusted R-squared: 0.7398
F-statistic: 33.21 on 3 and 31 DF, p-value: 8.314e-10
```

For clarity, we have omitted the output about usually large residuals, or the significance codes. Polynomial contrasts test for linear (L), quadratic (Q) and cubic (C) shapes of the relationship between the factor levels and the means for the groups. Here it is obvious that the relationship is linear, as we suspected: there is no evidence of any curvature (quadratic or cubic) to the relationship.

Using contrasts

Using our own contrasts, we first obtain the coefficients for all of our specific predictions (just as we did above for AQB), and tell R that they are going to be used as a set of contrasts for the factor temperature. They are entered as a matrix using cbind (bind by columns); the c() function combines numbers into a vector:

```
> contrasts(temperature) <- cbind(c(-3,1,1,1),c(0,-2,1,1),c(0,0,-1,1))
```

We can check that we have done this correctly by typing:

```
> contrasts(temperature)
```

```
    [,1] [,2] [,3]
10   -3    0    0
15    1   -2    0
20    1    1   -1
25    1    1    1
```

and we see that R knows what the factor levels are called, and has the correct pattern of coefficients (the columns). Note that each column must sum to zero, by definition. The product of any two columns must also sum to zero (if they do, this means the two contrasts are independent of one another: they are 'orthogonal' in the jargon).

Then we ask R to do the same ANOVA, but give us the summary of the linear model that uses the contrasts we have provided:

```
> summary.lm(aov(light ~ temperature))
```

```
aov(formula = light ~ temperature)
Coefficients:
              Estimate Std. Error t value Pr(>|t|)
(Intercept)   160.1679     0.9996 160.229  < 2e-16 ***
temperature1    4.8655     0.6070   8.015 4.74e-09 ***
temperature2    4.7167     0.7653   6.163 7.72e-07 ***
temperature3    5.0833     1.4239   3.570  0.00119 **
Residual standard error: 5.696 on 31 degrees of freedom
Multiple R-squared: 0.7627,    Adjusted R-squared: 0.7398
F-statistic: 33.21 on 3 and 31 DF, p-value: 8.314e-10
```

Again, for clarity, we have omitted some output.

In the table, the 'Intercept' estimate is the overall mean of the data. Each contrast estimate is complicated (see Crawley, 2007: 372–374 for details), made even more so here by the unequal sample sizes. Thus these results differ from those of AQB because R interprets the contrasts differently, and is adjusting for the different sample sizes in a sophisticated way, but the conclusions are the same.

Notice that all these ANOVAs get the same value for F. They divide the groups in different ways, but the same four group means are always involved.

The conclusion from both AQB and R is that there is good evidence that temperature increases light absorbance, and hence also the rate of reaction. The effect is linear, at least within the range of temperatures studied here.

Box 3.5c Mean ranks (medians): a *general* non-parametric one-way ANOVA for two or more groups (Kruskal–Wallis test)

An experimenter measured how much alcohol students in all-female (F), mixed (MF), and all-male (M) halls of residence drank in one month. The data for the variable being analysed (*Alcohol*, the units of alcohol consumed in one month by a student) depart significantly from a normal distribution (Shapiro–Wilk = 0.96, d.f. = 300, $p < 0.001$), so we can opt to use a non-parametric test. The grouping factor, *Hall type*, has three levels. The experimenter's prediction was a general one; she merely asked whether these types of hall of residence differed in their alcohol consumption per student, so that the expectation was that F ≠ MF ≠ M.

A general non-parametric one-way ANOVA in AQB
In AQB, choose the 'np 1-way' worksheet and enter the data in groups, one group per column (Fig. (i)). This option will perform a Kruskal–Wallis non-parametric one-way ANOVA. Clicking the 'go' button in the 'RESULTS' box to the right leads to the out-

come being displayed in the box. H is the Kruskal–Wallis test statistic, which here has two degrees of freedom (the number of groups minus one), so there is good evidence of a difference between the halls in alcohol consumption ($H = 104.6$, d.f. = 2, $p < 0.001$).

A general non-parametric one-way ANOVA in R
As usual, in R the data (alcohol) form one column, and the type of hall (hall) another column. After reading this into R, check whether R knows hall is a factor:

```
> is.factor(hall)
[1] TRUE
```

so that's fine. Now we plot the medians in a box plot (Fig. (ii)):

```
> plot(hall, alcohol, ylab=c("units
  of alcohol"))
```

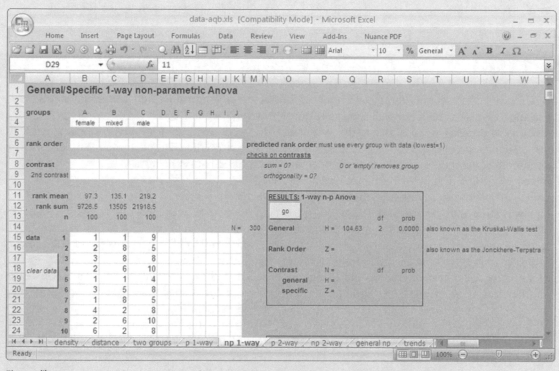

Figure (i) Data entry and results output for a general non-parametric one-way ANOVA in AQB.

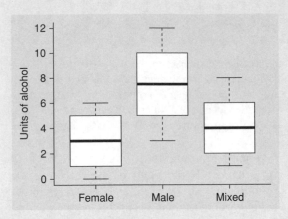

Figure (ii) Median (with interquartile ranges and extremes) number of units of alcohol drunk in one month by students living in three types of halls of residence (all-female, all-male, or mixed).

Notice how R orders the levels alphabetically rather than in the order of input. It looks from the plot that male halls drink quite a bit more. To perform the test, we merely type:

```
> kruskal.test(alcohol, hall)
```

or alternatively

```
> kruskal.test(alcohol ~ hall)

Kruskal-Wallis rank sum test
data: alcohol and hall
Kruskal-Wallis chi-squared =
104.6252, df = 2, p-value < 2.2e-16
```

Note that R calls the test statistic a χ^2 instead of H; this is because as sample size increases, the distribution of H comes to approximate that of χ^2 – they thus refer to the same thing here (hence the value from R is the same as that from AQB).

The conclusion is therefore that there is good evidence of differences in alcohol consumption among halls (H (or χ^2) = 104.6, d.f. = 2, $p < 0.001$).

Post-hoc **multiple comparisons**

There is no generally accepted *post hoc* multiple comparisons test for non-parametric analyses of variance, and few if any are implemented in the standard statistical packages. However, they do exist. The R package npmc has been written specifically to do them. If you want to try them, you should download and install the package from the R website.

Box 3.5d   Mean ranks (medians): a *specific* non-parametric one-way ANOVA

Again, we can use non-parametric analysis of variance to test specific as well as general predictions of difference. To illustrate this, we shall use a similar data set to that of Box 3.5c, on the amount of alcohol drunk by students at all-female, mixed or all-male halls of residence during the course of one week. Once again, the residuals are highly non-normal (Shapiro–Wilk = 0.96, d.f. = 300, $p < 0.001$), and so a non-parametric test is again appropriate.

Since it is well known that young men are more likely to be heavy drinkers, and to egg one another on in a drinking environment, than women, an obvious specific prediction is that all-female halls (A) will have lower alcohol consumption than mixed-sex halls (B), which, in turn, will have lower consumption than all-male halls (C).

A specific non-parametric one-way analysis of variance in AQB

In AQB we can test this prediction explicitly using a predicted rank order. We can also test it indirectly using contrasts.

Testing rank order

In the 'np 1-way' worksheet again, enter the groups and their predicted rank order in the 'groups' and 'rank order' boxes (Fig. (i)) and click 'go' in the 'RESULTS' box to the right. You can see that the probability is quite a bit lower for the test for a specific rank order (0.0014) than for the general test for differences from Box 3.5c (0.0058). This demonstrates that the specific test, if the prediction is supported by the data, is much more powerful than the equivalent general test.

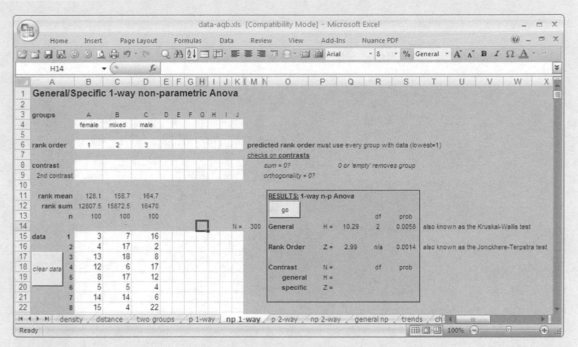

Figure (i) Data entry, rank order prediction and results output for a non-parametric specific one-way ANOVA in AQB.

Using contrasts

We can also test the specific prediction using contrasts, since we can (see Box 3.5b), with our two degrees of freedom, predict that:

$$A < (B + C)/2, \text{ i.e. } 2A < (B + C),$$
$$\text{or } 0 < -2(A) + 1(B) + 1(C)$$
and $\quad B < C, \text{ i.e. } 0 < -1(B) + 1(C)$

Entering the first set of contrast coefficients into the 'contrasts' box in AQB (Fig. (ii)) and clicking 'Go' gives the outcome in the 'RESULTS' box. We can see that the contrast is significant ($z = 3.17$, $p < 0.001$). Repeating this for the second set of coefficients shows that although the contrast is in the right direction, it is not significant ($z = 0.62$, ns).

A specific non-parametric one-way analysis of variance in R

We input the data as usual, with all the alcohol values in one column, and the hall names in another.

Since we will be predicting an order of expected treatment mean ranks, we need to be careful about how R orders the treatments. It does so alphabetically, whatever order you have put them in. You can find out what R thinks the order is by typing:

```
> levels(hall)
[1] "female" "male" "mixed"
```

You can see this is not the order we want them in.

We can either use this order for our predicted mean ranks, or we can make R use a sensible order by typing:

```
> hall <- ordered(hall,levels=
  c("female","mixed","male"))
```

Once again we plot the medians in a boxplot (or the means ± s.e.):

```
> plot(hall,alcohol,ylab=c("units of
  alcohol"))
```

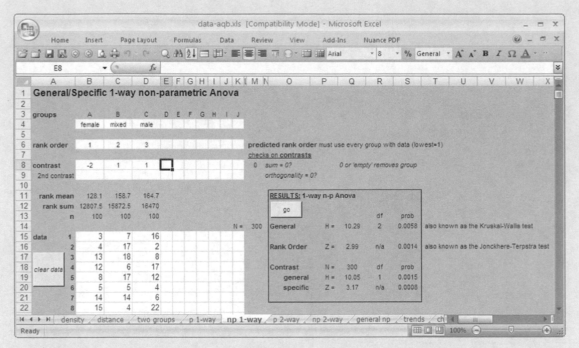

Figure (ii) Use of contrasts and results output for a non-parametric specific one-way ANOVA in AQB.

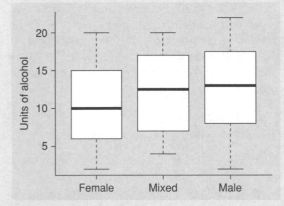

Figure (iii) Median (with interquartile ranges and extremes) number of units of alcohol drunk in one week by students living in three types of halls of residence (all-female, all-male, or mixed).

The medians certainly fall into the predicted pattern in this data set (Fig. (iii)).

Testing rank order
There is no built-in routine for this test, and so we will have to write one. It should be fairly

obvious how the commands relate to the formula (given in Appendix II, Box A3b(ii)). The final 'Z' prints the result to the screen:

```
> mds <- function(data,grps,coeff) {
r1<-rank(data)
rsum<-tapply(r1,grps,sum)
nsum<-tapply(r1,grps,length)
L<-sum(coeff*rsum)
N<-length(data)
E<-(N+1)*sum(nsum*coeff)/2
V<-(N+1)*(N*sum(nsum*coeff^2) - sum
   (nsum*coeff)^2)/12
Z<-(L-E)/sqrt(V)
Z
}
```

Note that for simplicity this routine does not take into account the correction for tied scores – we shall incorporate this in the routine in Box 3.9d. Thus the routine here will be slightly wrong when there are tied ranks, but only substantially wrong in the face of a large proportion of the data consisting of tied ranks.

We then type in the expected order of the rank means. Here we have reordered the levels as outlined above, so the expected order is 1,2,3: then we invoke our new function:

```
> c1 <- c(1,2,3)
> mds(alc, hall, c1)
[1] 2.985447
```

From both AQB and R, we conclude that the data on alcohol consumption among halls of residence are consistent with the pattern of female < mixed < male halls ($z = 2.99$, $p < 0.01$).

Using contrasts

There is no built-in routine for this test either in R, so again we will have to write one. This is very straightforward once we have given the vector of contrasts to R. First we create a vector the same length as the data with the appropriate contrast coefficients filling the vector elements. Then we simply create two vectors from the data, selecting first for negative and then for positive coefficients. This lumps all the groups on one side of the contrast together, and similarly for the other side. Then we simply carry out a Wilcoxon (Mann–Whitney) one-tailed test:

```
npc<- function(data,grps,contrasts) {
contr<-rep(contrasts,table(grps))
dneg<-data[contr<0]
```

```
dpos<-data[contr>0]
wilcox.test(dneg,dpos,alternative=
  c("less"))
}
```

We now specify the contrast, and invoke the function:

```
> c1 <- c(-2,1,1)
> npc(alc,hall,c1)

Wilcoxon rank sum test with
continuity correction
data: dneg and dpos
W = 7757.5, p-value = 0.0007616
alternative hypothesis:
true location shift is less than 0
```

and the second contrast:

```
> c1 <- c(0,-1,1)
> npc(alc,hall,c1)

Wilcoxon rank sum test with
continuity correction
data: dneg and dpos
W = 4746, p-value = 0.2675
alternative hypothesis:
true location shift is less than 0
```

As with AQB, the predictions are in the correct order, but only the first one (female vs mixed + male) is significant.

Box 3.5e   A parametric repeated-measures ANOVA

A molecular biologist wanted to study the impact of two different selection regimes on the production of oxygen free radicals. He therefore distributed one culture of liver cells (which he obtained from various sources) to 100 of his colleagues – a different culture going to each – and asked them to divide each into three selection treatments (randomly selected control, selected for high growth rate, and selected for low growth rate). After 10 rounds of cell division under these conditions, his colleagues sent back their cultures and the molecular biologist measured the average production of free radicals of 500 cells (a measure of metabolic activity) of each treatment from each culture.

This is a one-way repeated-measures design because cells from each culture are exposed to all treatments, and there is only one sort of treatment (selection type). Cultures may differ in lots of ways from each other, including their natural level of free radicals; 'culture' is therefore a block effect that needs to be allowed for in the analysis in order to see the impact of the treatments. The researcher only obtained a single measurement of the dependent variable (level of free radicals) from each culture and each treatment.

First we frame the prediction: that the selection regimes change the level of free radicals from the randomly selected control condition (H_1), without specifying the direction of change (this is therefore a general hypothesis). As with the usual parametric one-way ANOVA (see Box 3.5b), it is not possible to test a specific version of this hypothesis, although, as before in AQB, one can approach it using contrasts. The null hypothesis is that selection does not change the level of free radicals (H_0).

Repeated-measures ANOVA using AQB

Select the 'p 1-way' sheet in AQB and enter the measurement for each treatment as in Fig. (i), with the data for each culture on a single row. Therefore there will be equal numbers of values in each column (treatment), and if this is so, the analysis will automatically appear in the RESULTS box labelled '1-way repeated-measures ANOVA'. There are two F-values. The lower one is a test for the significance of differences among the 'blocks', i.e. the repeated-measures factor. Here it is huge and highly significant ($F = 1943.0$, d.f. $= 99,198$, $p < 0.001$). The overall mean values for each of the columns are quite similar (about 68–70) with a fairly high SE (ca. 2.5). However, once this block effect is allowed for, the treatment effect (the upper F-value) is also significant ($F = 83.3$, d.f. $= 2,198$, $p < 0.001$).

In the RESULTS box for a simple ANOVA (i.e. *not* repeated measures), which also appears automatically, you can see that if there had been no repeated-measures factor, then the large variation within the columns would obscure any treatment effect.

Figure (i) A general parametric repeated-measures ANOVA in AQB.

Repeated measures ANOVA using R

In R the data are in the usual format of a single column of the data (the reactive oxygen species measurements, called `ros`), and a single column (`treat`) indexing the three treatments as names ('random', 'high' and 'low'). Check that `treat` is a factor:

```
> is.factor(treat)
[1] FALSE
```

If it says 'FALSE' then make it a factor:

```
> treat <- factor(treat)
```

Now we must create another factor (`cult`) for the blocks:

```
> cult <- factor(rep(c(1:100),times=3)
```

We cannot plot mean values that take account of the repeated measures, but we can just plot the groups means to see what they look like. Obtain the means (`tmeans`) and the SEs (`tses`) in the usual way, and plot them using the routine of Box 2.3 (Fig. (ii)). There do not seem to be any differences worth speaking about.

Then we run an ANOVA using `treat` and `cult` as factors `cult` will take out the variation due to differences in cultures (allowing for inter-individual variation) from the residual error, leaving the variation due to treatment to be evaluated against the remaining error variation. As we shall see in Box 3.6, having two factors as predictors potentially allows an interaction term: here there is no interaction, just the main effects.

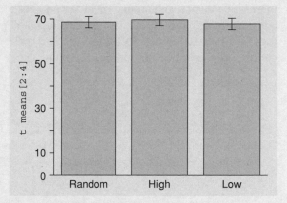

Figure (ii) Mean (± s.e.) production of free radicals (mol s^{-1}) of 500 cultured liver cells under three selection regimes (control, selected for high growth rate, and selected for low growth rate).

```
> m1 <- aov(ros ~ treat + cult)
> summary(m1)
          Df  Sum Sq Mean Sq  F value     Pr(>F)
treat      2     165   82.75   83.293  < 2.2e-16 ***
cult      99  191113 1930.44 1943.166  < 2.2e-16 ***
Residuals 198    197    0.99
```

As in AQB, the treatment effect is highly significant. A test shows that the residuals are not significantly different from a normal distribution:

```
> shapiro.test(resid(m1))

Shapiro-Wilk normality test
data: resid(m1)
W = 0.9923, p-value = 0.1212
```

Notice here the much greater power of the repeated-measures design in cases like this where there is large variation among individuals. By removing the large inter-culture variation, the impact of the treatments becomes evident over and above variation among individual cultures. There are no obvious differences among treatments from the means ± s.e., yet the effects are highly significant.

Box 3.5f   A non-parametric repeated-measures ANOVA

Suppose a microbiologist was interested in the impact of three amino acids (leucine, alanine and lysine) on cultures of *Aspergillus* fungi. She knew from the literature that alanine and especially lysine should stimulate the production of colonies. However, she had only a very small incubator so that she could experiment with only three Petri dishes at a time. This was a problem, because each time she performed the experiment there was uncontrollable variation in culture ages, temperatures, concentrations, etc. She therefore added each amino acid to one dish and incubated the three dishes, but repeated this 50 times to obtain 50 replicates. She then counted the number of colonies on each plate.

Each set of three constitutes the repeated measures, or block.

A non-parametric repeated-measures ANOVA using AQB

Because of prior knowledge, we make the specific prediction that the rank order will be leucine < alanine < lysine (H_1). The null hypothesis (H_0) is that the mean ranks will not follow this pattern.

In AQB (Fig. (i)) the data for the treatments are in columns, with the data for each replicate on a single row. The hypothesis is placed as a predicted rank order in row 6, as shown, with the smallest predicted rank mean given the rank of

Figure (i) A specific non-parametric repeated-measures ANOVA in AQB.

'1'. Clicking 'Go' in the RESULTS table for repeated-measures produces the result. Both the general tests (H) and the specific one (z) are produced. It is clear that the result is very significant ($z = 2.90, p < 0.01$), indicating agreement with the prediction.

Clicking 'Go' on the upper RESULTS panel, the non-repeated-measures design, shows that without taking the block factor into account, this experiment would have concluded that there were no differences of any kind among these treatments.

A non-parametric repeated-measures ANOVA using R

The data are read in the usual way, data in one column (`fungus`), and treatments in another (`aa`). We need an index to the blocks (i.e. the rows) as well:

```
> blk <- rep(c(1:50),times=3)
```

The medians can be plotted using the usual plot function, but as with the parametric case, this takes no account of the repeated measures, and can be very misleading. As Fig. (ii) shows, there do not seem to be any differences among the treatments here. The test goes by the name of the Friedman test in R and many packages:

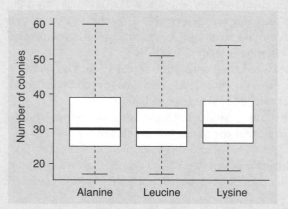

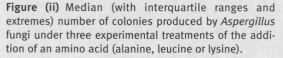

Figure (ii) Median (with interquartile ranges and extremes) number of colonies produced by *Aspergillus* fungi under three experimental treatments of the addition of an amino acid (alanine, leucine or lysine).

```
> friedman.test(fungus, aa, blk)

Friedman rank sum test
data: fungus, aa and blk
Friedman chi-squared = 10.4385,
df = 2, p-value = 0.005411
```

To perform the specific version of this test will require us to write the routine ourselves. The data are better organised as a dataframe (`dtfr`) of three columns, so that the repeated measures lie on a single row. The routine is straightforward except for the correction for tied ranks, common in data sets consisting of integer numbers. The first eight lines rank the data within blocks, and then sum these ranks for each treatment (`rsums`); these are then used in the basic calculation of the Z-test, involving the observed (`OL`), expected (`EL`) and variance (`VARL`) of the rank means. The tricky bit is the correction for ties (`tcorr`) within each block:

```
> nprm.s <- function(dtfr,contrasts) {
cols<-length(dtfr[1,])
rows<-length(dtfr[,1])
len<-rows*cols
ranks<-c(1:len)
ranks<-matrix(ranks,nrow=rows)
for (i in 1:rows)
  ranks[i,]<-rank(dtfr[i,])
rsums<-colSums(ranks)
OL <- sum(rsums*contrasts)
EL<-0
VARL<-0
tie<-0
btie<-0
E1 <- sum(contrasts*contrasts)
E2 <- sum(contrasts)
for (i in 1:rows) {
EL = EL + ((cols+1)*E2/2)
btie=tie
jj<-cols-1
for (j in 1:jj) {
kk=j+1
for (k in kk:cols) {
if (dtfr[i,j]==dtfr[i,k]) tie=tie+1
  else tie=tie
}}
```

```
if (btie<tie) nt=tie-btie else nt=0
if (nt==1) nt=nt+1
tcorr = (1-(((nt^3)-nt)/((cols^3)
  -cols)))
VARL = VARL + (cols+1)*((cols*E1)
  - (E2*E2))*tcorr/12
}
Z = (OL - EL)/sqrt(VARL)
Z
}
```

We then enter the predicted contrasts, and invoke the routine:

```
> c1 <- c(1,2,3)
> nprm.s(d1,c1)
[1] 2.895691
```

The significance of this z-value is given by:

```
> 1-pnorm(2.895691)
[1] 0.001891624
```

As before, this is a very significant difference.

Covariates. Sometimes it is useful to control for other, nuisance, factors that are measured on a constant-interval scale within an analysis of variance. For instance, an analysis of differences in mating success between territory-owning and non-territory-owning male wood mice (*Apodemus sylvaticus*) might want to control for body size to rule out an effect of territory ownership arising simply because owners tend to be bigger. In a parametric analysis of variance, body size could be incorporated as a *covariate*. The analysis would then reveal the independent effects of territory ownership and body size. Most major statistical packages allow covariates to be included in analyses of variance. Box 3.6 shows how to do this in R.

Box 3.6   Analysis of covariance – taking a covariate into account

A physiologist was interested in the impact of three foods on the basal metabolic rate (BMR) of rats, and conducted an experiment in his laboratory where rats in individual cages were randomly assigned to one of the three foods (A, B and C), and were fed this food over one week. He then measured the BMR of each rat three times, taking the average as the value for each rat. He suspected, however, that body size might also influence BMR, and might obscure the result of the experiment unless controlled for. Thus he also recorded the weight of each rat when he measured its BMR.

The dependent variable for analysis is BMR, measured on a constant interval scale. The question concerns differences in the mean value of BMR between the three levels of the factor food type (*food*, after having taken into account the effect of a *covariate*, body size (*size* in Fig. (i)).

Frame the prediction. In this case the physiologist did not have any preconceptions about what to expect, and therefore just tested the general prediction that the mean values differed.

The analysis is called an analysis of covariance (ANCOVA), and this example is the simplest possible kind – a one-way analysis of variance with a single covariate. ANCOVA can be very complex, involving several factors and covariates, and hence it cannot be done using the simple AQB package. Thus we shall show how to do the analysis in R.

Analysis of covariance in R
In R the data are held in the usual column format, with a column for `bmr`, one for `size` and one for the factor `food`. Having read them into R, check that `food` is a factor:

```
> is.factor(food)
[1] TRUE
```

Plot the data to see what they look like (Fig. (i)). The `plot` command plots the data, using a different symbol for each group (by converting the group names into numbers with `as.numeric`, and adding 16, the default character). Then the three lines of best fit are added using the `abline` command and dashed lines (`lty=2`):

```
> plot(size,bmr,pch=16+as.
  numeric(group))
> abline(lm(bmr[food=="A"]
  ~size[food=="A"]),lty=2)
> abline(lm(bmr[food=="B"]
  ~size[food=="B"]),lty=2)
> abline(lm(bmr[food=="C"]
  ~size[food=="C"]),lty=2)
```

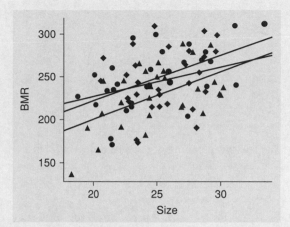

Figure (i) Basal metabolic rate (VO_2 g^{-1} s^{-1}) of rats fed three different foods (three regression lines, and different symbols: circles, food A; triangles, food B; diamonds, food C).

It certainly could be that each group is different, but there is a lot of scatter. If we just ask whether there are differences among `food` types in `bmr`, the answer is no:

```
> summary(aov(bmr~food))
          Df Sum Sq Mean Sq F value  Pr(>F)
food       2   6744  3372.1  2.6746 0.07467 .
Residuals 86 108430  1260.8
```

Looking at the plot (Fig. (i)), there is little evidence from the plot for lines with different slopes. We therefore fit a model of just a single line (covariate) for `size`, plus the differences among `food` types, using the `lm` command (meaning 'linear model'):

```
> m1 <- lm(bmr ~ size + food)
```

Be aware that the order matters. The result would be different if we put '`food + size`'. This is because we are first going to allow for the effect of size, and then ask whether residuals from that relationship differ among groups. This would be very different from first looking at differences in the mean values of groups, and then asking whether the residuals were related to size.

The ANOVA table appears when we type:

```
> anova(m1)
Analysis of Variance Table
Response: bmr
          Df Sum Sq Mean Sq F value    Pr(>F)
size       1  23068 23068.1  22.924 7.05e-06 ***
food       2   6573  3286.5   3.266    0.043 *
Residuals 85  85534  1006.3
```

Having allowed for `size` effects, we see that there are indeed significant differences among the `food` types ($F_{2,85} = 3.27$, $p < 0.05$).

We can check whether we were correct in only fitting a single line for the effect of size: perhaps the slope of this relationship is different for each of the food groups. This is fitted as an interaction between `size` and `food` – we will learn more about interactions shortly:

```
> anova(lm(bmr~size + food + size:food))
Analysis of Variance Table
Response: bmr
          Df Sum Sq Mean Sq F value    Pr(>F)
size       1  23068 23068.1 22.5470 8.47e-06 ***
food       2   6573  3286.5  3.2122   0.04532 *
size:food  2    615   307.7  0.3008   0.74106
Residuals 83  84918  1023.1
```

We see there is no evidence of different slopes among groups.

1 × n chi-squared. One-way analysis of variance uses a comparison of mean (or mean rank) values of individual data values to arrive at a test statistic. Where data values are counts, however, we could, as in the two-group case (*see* Box 3.2) perform a chi-squared test on the *totals* for each group. We then have what is known as a $1 \times n$ chi-squared analysis, where n is the number of groups. The two-group (1×2) chi-squared test earlier is just one form of this, and the calculation of expected values and the χ^2 test statistic are exactly the same as for the two-group case. Box 3.7 shows how to do these in AQB and R.

Tests for differences in relation to two levels of grouping

In all the above difference tests, we were concerned with differences within a single level of grouping, e.g. between groupings based on seed colour. However many groups of seed colour we had (red, yellow, green, orange, blue, etc.) we would still be dealing only with seed colour and thus with one level of grouping. But we can easily envisage situations in which we would be interested in more than one level of grouping. For example, we might want to know not only whether chicks peck at some colours of grain more than others, but also whether pecking in males is different from that in females. More interestingly still, we might want to know whether the sex of chicks affects the *difference* in pecking at the various colours of grain. Is the difference stronger in one sex? Is it in the same direction in both sexes? Here we have *two* levels of groupings: seed colour and sex. In the examples that follow, we shall look at analyses that cater for two levels of grouping with two groups in each.

2 × 2 chi-squared. If we have data in the form of counts, we can again use chi-squared, but the expected values in a 2×2 analysis (and in any other $n \times n$ chi-squared analysis) are calculated in a different way from those in a $1 \times n$ analysis. Instead of taking equal values, or values determined on the basis of some a priori expectation, the rows and columns of the 2×2 (or $n \times n$) table are totalled and the grand total calculated, and the respective expected values for each cell of the table are calculated as (row total × column total)/grand total. Box 3.8 shows the procedure for performing a 2×2 chi-squared analysis in AQB and R. There is a very clear account of how the expected values work in Crawley (2007: 301–304).

Box 3.7 A test comparing counts classified into two or more groups ($1 \times n \, \chi^2$ test)

A clinical microbiologist was assaying the effect of four antibiotics on a bacterial culture. To see whether the antibiotics differed in their ability to kill the bacterium, he counted the number of cultures on which clear plaques appeared after drop treatment with each one, and performed a χ^2 test, assuming equal expected values across the four antibiotics. The data

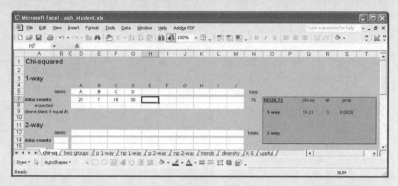

Figure (i) Performing a $1 \times n \, \chi^2$ test in AQB.

consist of the total numbers of cultures. The nominal grouping factor is *treatment*, with four levels (the different antibiotics).

Frame the prediction, here that antibiotics differ in their ability to kill the bacterium, and hence there will be differences in the numbers of plaques between antibiotics (H_1). The null hypothesis is that all the antibiotics are equally effective, and hence there will be no differences (H_0).

Calculating a $1 \times n \, \chi^2$ using AQB

Under the 'chi-sq' worksheet in AQB, the counts for each treatment are entered as in Fig. (i). Leaving the 'expected' values empty is the default assumption of equal expected numbers. The result appears automatically in the 'RESULTS' box to

the right, showing, in this case, a significant difference between antibiotics ($\chi^2 = 14.21$, d.f. = 3, $p < 0.01$).

Calculating a $1 \times n$ chi-squared using R

To test the total counts, we use the routine as for Box 3.2, but here for four groups instead of two. The default null hypothesis is for the numbers to be the same:

```
> chisq.test(c(21,7,18,30))

Chi-squared test for given
probabilities
data: c(21, 7, 18, 30)
X-squared = 14.2105, df = 3, p-value
= 0.002632
```

2 × 2 two-way analysis of variance. As before, if we want to compare sets of individual data values, we can use analysis of variance, but this time it is a two-way rather than a one-way analysis. In a 2 × 2 two-way analysis of variance, the data are cast into four cells (two × two groups, which can be cast as two rows and two columns of a table). If we wanted to do a 3 × 5 two-way analysis the data would be cast into 15 cells, and so on for any combination of levels of grouping. Once again there are both parametric and non-parametric versions of the analysis (Boxes 3.9a–d). As usual, the parametric test assumes the data conform reasonably to normality. This assumption is, of course, relaxed for the non-parametric equivalent. However, *both* parametric and non-parametric tests assume that the data have the same variance within each cell (i.e. within each combination of levels of grouping) – another example of the distribution-free, but not assumption-free, nature of non-parametric tests. Both types of analysis compare

Box 3.8   A test comparing counts in a 2 × 2 classification (2 × 2 χ^2 test)

A count was made of the total number of men and women with stomach biopsies showing a presence or absence of cancerous cells. We have two grouping factors, each with two levels: *sex* (men/women) and *cancer* (with/without), and each person's result is classified into one of the four combinations.

The prediction is that the sex of the person affects their chance of showing stomach cancer (H_1); in a 2 × 2 chi-squared this would be reflected as a significant non-independence between the two factors (*see below*). The null hypothesis is that sex has no effect on the incidence of stomach cancer (H_0).

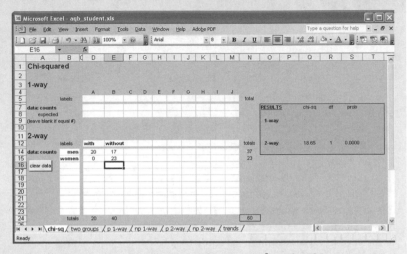

Figure (i) Data entry and results output for a 2 × 2 χ^2 test in AQB.

A 2 × 2 χ^2 test in AQB

Under the 'chi-sq' worksheet in AQB, simply enter the group labels and four totals into the cells of the '2-way' box, as in Fig. (i), and the outcome appears automatically in the lower, two-way, line of the 'RESULTS' box. So, in the example here, there is a significant interaction between sex and cancer incidence (χ^2 = 18.65, d.f. = 1, $p < 0.001$).

A 2 × 2 χ^2 test in R

In R, we simply declare the numbers to be in matrix form, and then R assumes this is a two-way chi-squared test. The numbers are read in column-wise, so don't get confused!

```
> cancers <- matrix(c(20,0,17,23),
  nrow=2)
> cancers
     [,1] [,2]
[1,]   20   17
[2,]    0   23
```

Then we invoke the test by typing:

```
> chisq.test(cancers)

Pearson's Chi-squared test with
Yates' continuity correction
data: cancers
X-squared = 16.2955, df = 1, p-value
= 5.419e-05
```

Notice that this is not the same result as for AQB, although it is also highly significant. This is because AQB implements a simple rule for applying a correction for low numbers (the Yates' correction). In AQB the rule is that it is applied if the expected frequency is less than 5. This is a simplified version of a more accurate but more complicated rule implemented in statistical packages. You can see that R has applied the correction whereas AQB has not, by forcing R not to use it:

```
> chisq.test(cancers, correct = F)

Pearson's Chi-squared test
data: cancers
X-squared = 18.6486, df = 1,
p-value = 1.572e-05
```

and we get the value that AQB provides.

The conclusion from these analyses is that the two factors interact, i.e. the chance of getting stomach cancer differs between men and women ($\chi^2 = 16.3$, d.f. = 1, $p < 0.001$).

the mean values of the columns within the classification and the mean values of the rows. In other words they compare means within each of the two levels of grouping. Comparisons between the column means or between the row means are known as the *main effects* and are distinguished from a second kind of comparison referred to as an *interaction*. (Note that an interaction can be calculated only if all cells contain more than one replicate value.) If there is a significant interaction it means the two sets of samples at one level of grouping respond differently to differences in the second level of grouping. An example makes the distinction between main effects and interaction clear. Imagine our two levels of grouping are freshwater versus marine fish and male versus female, and the variable for comparison is growth rate. Freshwater versus marine can be the columns of the classification (*see table below*) and male versus female the columns. The analysis would be concerned with the following: (a) main effect 1: differences in column means (is there any difference in growth rate between freshwater and marine fish?), (b) main effect 2: differences in row means (is there any difference in growth rate between males and females?) and (c) any inter-action between the levels of grouping (e.g. is the difference in growth rate between males and females greater in one environment than in the other?).

As in the one-way analysis of variance, the non-parametric version can be used to test either general or specific predictions, but now about both the two main effects and any interactions. Making specific predictions is rather more involved in the two-way analysis because we need to be clear as to exactly what we are comparing and to calculate different coefficients for each specific comparison so predictions can be tested. Box 3.9c illustrates the procedure.

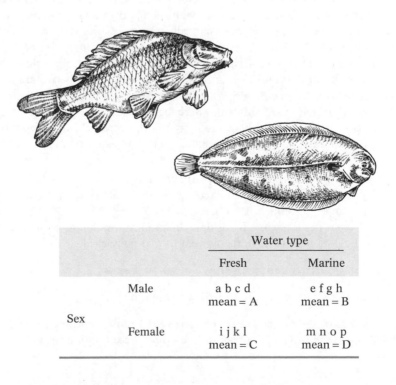

		Water type	
		Fresh	Marine
Sex	Male	a b c d mean = A	e f g h mean = B
	Female	i j k l mean = C	m n o p mean = D

Box 3.9a Mean values – a *general* parametric two-way ANOVA

An experiment aimed to detect the effects of food type and parasitic infection on an enzyme (liver alcohol dehydrogenase) in rats (*Rattus norvegicus*). The variable being analysed (*enzyme*) is a measure of the enzyme's rate of reaction, and there are two factors, each with two levels: *food* (peas, beans) and *parasitism* treatment (unparasitised, parasitised). The experimental design required each rat to be in one food group (peas or beans) and one parasitism treatment group (unparasitised or parasitised). Testing the residuals for normality (see Box 3.1b) showed that the data could be assumed to be normal (Shapiro–Wilk = 0.99, d.f. = 160, ns).

Because of the two-way design, we have three predictions, one about the effects of *food*, one about the effects of *parasitism* (together called the main effects), and one about the *interaction* between food type and parasitism (e.g. is the effect of food type affected by whether or not the rat is also parasitised?)

Doing a general parametric two-way ANOVA in AQB

In AQB, select the 'p 2-way' worksheet and enter the data as usual in columns, making sure that they are in the correct ones. The advantages of intuitive inputting of the data in columns for one-way designs are lost for two-way designs, and you must be careful.

It is often helpful to draw a diagram of the design, as in Fig. (i)a. This shows the two levels of *food* as the columns, and the two levels of *parasitism* as the rows. In the cells of this design in Fig. (i)a, we have written the column in which the data for that cell should be placed in AQB, and indicated on Fig. (i)b by arrows where the data should go. On row 3 of the worksheet are spaces for the factor labels – in our case, Factor A is *parasitism*, and factor B is *food*. There is space on the worksheet for four levels of Factor A and six levels of Factor B – these are indexed by number on rows 4 and 5 of the worksheet, and have spaces for the level labels on rows 6 and 7, respectively. Thus columns B–G are for the six possible levels of Factor B for level 1 of Factor A; columns H–M are for the six possible levels of Factor B for level 2 of Factor A; and so on up to level 4 of Factor A. Rows 9 and 10 of the worksheet are for the contrasts, as we have met before and will meet in Box 3.9b.

Then come five rows which automatically display (as you type in the data) the summary statistics of the data: rows 12, 13 and 14 record the mean, s.e. and sample size of the data for each column (i.e. for each cell of the two-way design). Row 15 has four boxes, which show the mean value for each of the levels of Factor A (*parasitism* in our design, and hence only two are filled in, since we have only two levels for this factor). Finally, row 16 has six boxes for the mean values of the six possible levels of Factor B (but only two are filled in here, because *food* has only two levels).

There is room for a 4 × 6 two-way design in the spreadsheet, but anything larger will need to be done in R or another package. The result automatically appears in the RESULTS box (Fig. (i)c).

Doing a general parametric two-way ANOVA in R

In R the data are imported from an Excel tab-delimited file in the usual way, with all the `enzyme` activity values in a single column, with other columns coding the `parasitism` (parasitised, unparasitised) and `food` type (peas, beans). If we have them in four columns as in the AQB screen, then we can reformat the data in R to be in the correct format quite easily:

```
> enzyme <- c(p.peas,p.beans,u.peas,u.beans)
```

(or whatever your headings are called), and then the factors:

```
> parasitism <- factor(rep(c("parasitized","unparasitized"),each=80))
> food <- factor(rep(c("peas","beans"),each=40,times=2))
```

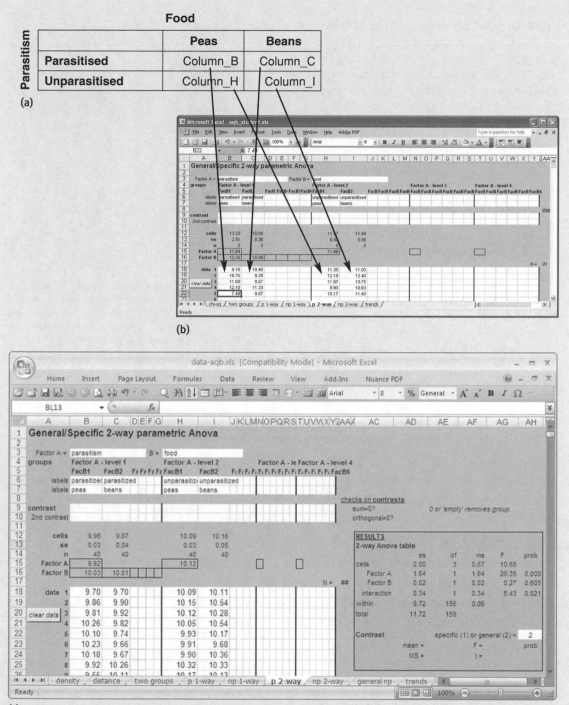

(c) Note that the width of some of the columns has been reduced to fit data and result into a single frame. Note carefully how the two-way design is put into the column format.

Figure (i) (a) Experimental design, (b) data entry and (c) results output for a general parametric two-way ANOVA using AQB.

Make sure the indexing of the values is correct!

We can plot the mean values (Fig. (ii)) by typing the following, using `beside=T` to put the bars next to one another rather than on top of one another. We could modify the `plot.error.bars` routine to add error bars, if desired. Thus:

```
> barplot(tapply(enzyme,
  list(parasitism,food),mean),
  beside=T)
```

The differences that exist look very small! Let's do the test, first performing the usual tests for normality and heterogeneity of variances:

```
> m1 <- aov(enzyme ~ parasitism *
  food)
> shapiro.test(resid(m1))
Shapiro-Wilk normality test
data: resid(m1)
W = 0.989, p-value = 0.2459
> fligner.test(enzyme ~ parasitism *
  food)
Fligner-Killeen test of homogeneity of variances
data: enzyme by parasitism by food
Fligner-Killeen:med chi-squared = 0.0562, df = 1, p-value = 0.8126
```

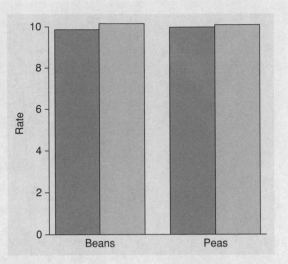

Figure (ii) Mean rate of reaction of liver alcohol dehydrogenase (mol s^{-1}) in rats allocated to treatments in a two-way design of food (peas or beans) and *parasitism* (unparasitised, parasitised). Dark bars indicated parasitised rats; light bars indicate unparasitised rats.

The diagnostics look OK, so we can go ahead and look at the ANOVA summary:

```
>summary(m1)
summary()
                Df Sum Sq Mean Sq F value    Pr(>F)
parasitism       1 1.6504 1.65039 26.4377 8.053e-07 ***
food             1 0.0166 0.01661  0.2660   0.60675
parasitism:food  1 0.3358 0.33581  5.3793   0.02167 *
Residuals      156 9.7384 0.06243
```

As with the AQB output, the main effect of the factor `parasitism` is highly significant, whereas that of `food` is not, but there is evidence of a significant interaction.

Note that the relevant degrees of freedom to cite with the *F*-ratio for each effect are the ones on the same line as the name of the effect in the table (the among-group d.f.), and in each case the error degrees of freedom (the within-groups d.f.).

From all this we conclude that:

■ there is evidence of differences in enzyme activity between parasitised and unparasitised animals ($F = 26.4$, d.f. = 1, 156, $p \ll 0.001$);
■ there is no evidence of an effect of food types on enzyme activity ($F = 0.27$, d.f. = 1, 156, ns);
■ there is evidence of an interaction between food type and parasitism that affects enzyme activity ($F = 5.38$, d.f. = 1, 156, $p < 0.05$).

We can present these statistical results either in the text of the Results section, or in the legends to the figures that display the mean values ± s.e.s, or we can reproduce the ANOVA table. They should be given only once, in one of these formats. It is good practice to give the table for all but the simplest of designs, because to experienced scientists it is very informative. When we are testing specific predictions within the ANOVA, the table becomes essential, as we shall see in Box 3.9b.

Fixed and random factors

We will just flag up here that there is an important difference between two types of factors that it is as well to be aware of. This is the distinction between fixed and random factors. All the factors in this book have been treated as fixed. The distinction between them is not absolute in many cases: it depends on what they represent in the experiment. The same factors can potentially be treated as fixed or random in different tests. The distinction really lies in the extent to which any significant difference emerging from an ANOVA can be generalised. Thus, for example, if food in our example is a fixed factor, and we find a significant difference between the levels of whatever we are measuring (here, enzyme activity), this implies only that there is a difference in the effect of peas and beans per se. If they are levels of a random factor, then it implies differences between any two randomly selected foods. A random factor therefore represents randomly selected levels of a universe of possible levels, and because they are chosen at random, then they represent the factor in general. Sometimes a factor is random because it is a block factor where we have no idea what the nature of the variation might be (e.g. differences among individuals, rearing cages, fields used for fertiliser plots, or physiological preparations), but it must be allowed for in the analysis. AQB is too simple a package to incorporate this distinction. However, whether a factor is fixed or random makes a big difference to the way the test statistic is calculated, and hence to the results, so it is important. Modern statistics uses special mixed-effects models where there are both fixed and random factors in the same analysis; they are beyond the scope of this book.

Box 3.9b   Mean values – a *specific* parametric two-way ANOVA

We have deliberately chosen a complex example here to show you how this works.

An experiment was designed to test the effect of temperature and motivation on the mathematical performance of students in a standard test. Students were tested one by one in an environmental chamber that could be set to one of three conditions of temperature (cold, normal warm room temperature or hot). Each student was pretreated in one of three ways to alter their motivation: (a) in a very off-hand manner and generally given the impression that they were incompetent (low motivation), (b) neutrally with respect to their competence (neutral) or (c) in a praising fashion, giving every encouragement (high motivation). Each student was randomly allocated to one temperature and one motivation treatment. Thus the target variable for analysis is mathematical performance, measured on a constant-interval scale. There are two grouping factors: *motivation*, with three levels (low, neutral, high) and *temperature*, with three levels (cold, warm, hot). We shall treat these as fixed factors (*see* Box above). Testing the residuals for normality (*see* Box 3.1b) demonstrated that the assumption of normality is acceptable (Shapiro–Wilk = 0.99, d.f. = 225, ns). The specific predictions deriving from this were:

- Prediction 1 (main effect 1): lower motivation will reduce mathematical performance,
- Prediction 2 (main effect 2): higher temperatures will reduce mathematical performance,
- Prediction 3 (interaction): higher temperatures and lower motivation together will reduce mathematical performance more than expected from their additive effects.

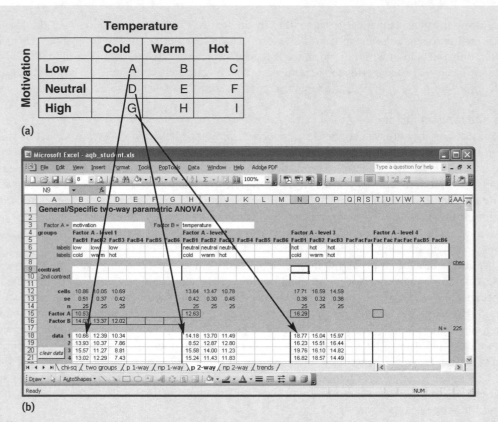

Figure (i) (a) The design and (b) data entry in AQB of the two-way ANOVA under consideration.

In order to see how to proceed, it is best to visualise the pattern of the design (Fig. (i)) and label each of the combination of treatments (the cells of the design matrix) with letters. As with the one-way ANOVA (Box 3.5b), we should, in theory, be able to test for a specified rank order of mean values, but, as we noted in Box 3.5b, this method has not been implemented in most statistical packages. We can test for a specified rank order in a non-parametric ANOVA (*see* Box 3.9c), but not easily in a parametric ANOVA. Once again, therefore, we shall use contrasts to test hypotheses that are almost the same thing.

Setting up the predictions

Prediction 1: We have three levels of the factor *motivation* (the rows in Fig. (i)), giving us two degrees of freedom, and hence two single-degree-of-freedom contrasts that we can use (as long as they are independent of one another). The first will contrast low motivation versus the other two, and the second will contrast neutral versus high. We proceed by casting them in terms of the cells of the design matrix (Fig. (i)):

(a) *high motivation > (neutral + low)*. The average performance for the high motivation cells of the design ($n = 3$) will be greater than those for the other cells ($n = 6$), i.e.:

$(G + H + I)/3 > ((D + E + F) + (A + B + C))/6$

which rearranges as:

$(-1)A + (-1)B + (-1)C + (-1)D + (-1)E + (-1)F + (2)G + (2)H + (2)I > 0$

where the numbers in the brackets are the coefficients of the first contrast (cf. Box 3.5b).

(b) *neutral > low motivation*. The average performance measured for neutral motivation ($n = 3$) will be higher than that for low motivation ($n = 3$), i.e.:

$(D + E + F) > (A + B + C)$

which rearranges as:

$(-1)A + (-1)B + (-1)C + (1)D + (1)E + (1)F + (0)G + (0)H + (0)I > 0$

where the numbers in the brackets are the coefficients of the second contrast.

Prediction 2: This is done in exactly the same way, involving the *temperature* factor (the columns of Fig. (i)) rather than *motivation* (the rows of Fig. (i)). The relevant coefficients are:

(a) *cold temperature versus other*. The average performance measured for the cold temperature cells of the design ($n = 3$) will be greater than that for the other cells ($n = 6$), i.e.:

$(A + D + G)/3 > ((B + E + H) + (C + F + I))/6$

which rearranges as:

$(+2)A + (-1)B + (-1)C + (2)D + (-1)E + (-1)F + (2)G + (-1)H + (-1)I > 0$

(b) *warm versus high temperature*. The average performance measured for the normal temperature cells of the design ($n = 3$) will be greater than that for the high temperature cells ($n = 3$), i.e.:

$(B + E + H)/3 > (C + F + I)/3$

which rearranges as:

$(0)A + (1)B + (-1)C + (0)D + (1)E + (-1)F + (0)G + (1)H + (-1)I > 0$

Prediction 3: This concerns the interaction between *temperature* and *motivation*, and it has four degrees of freedom (because the interaction has $(a - 1)(b - 1)$ degrees for the two factors A and B which have $a - 1$ and $b - 1$ degrees, respectively). We shall only use two of these degrees of freedom in casting the prediction into independent contrast sets. An interaction is *non-additivity of factors*, which means that the effect on one factor (say *temperature*) is different at the different levels of the other factor (*motivation*) (*see text*). This is complicated to visualise, and it is best to set out the interaction prediction as a difference between levels in the following way:

If the hypothesis is correct, then the *motivation* differences will increase as *temperature* decreases because both cold temperature and high motivation increase the ability of students to perform. The prediction is that this ability will increase more than the additive effect of each factor, thus:

(a) the motivation difference (*high* − (*neutral* + *low*)) will increase with temperature, so

$$cold > (warm + hot)$$

$$(G − (A + D)) / 3 > [(H − (B + E)) + (I − (C + F))] / 6$$

which rearranges as:

$$(−2)A + (1)B + (1)C + (−2)D + (1)E + (1)F + (2)G + (−1)H + (−1)I > 0$$

and:

(b) the motivation difference will follow the temperature pattern *warm* > *hot*

$$H − (B + E) > I − (C + F)$$

which rearranges as:

$$(0)A + (−1)B + (1)C + (0)D + (−1)E + (1)F + (0)G + (1)H + (−1)I > 0$$

Testing for specific differences in a two-way classification using a parametric two-way ANOVA in AQB

In AQB (Fig. (i)b), the data are as usual put into columns as indicated in the figure (*see* Box 3.9a for a description). Note the labels at the top that show how the two-way design is unpacked into a single set of adjacent columns (as designs become more complex, so the R method of data entry becomes more advantageous). The output for a general ANOVA appears automatically in the 'RESULTS' box (Fig. (ii)a), and you can see that the effect of Factor A (*motivation*) is significant (since $F = 160.65$, d.f. = 2,216, $p < 0.001$), as is Factor B (*temperature*) (since $F = 29.3$, d.f. = 2,216, $p < 0.001$), and there is a significant *motivation × temperature* interaction (since $F = 2.52$, d.f. = 4,216, $p < 0.05$).

Within the general ANOVA, we are testing specific hypotheses, since we expect higher temperature to *reduce* (rather than merely change) performance, and hence we enter '1' in the white box within the 'RESULTS' box (Fig. (ii)a). If the contrast coefficients sum correctly to zero, then entering them will bring up the result automatically. You can tell if they are correct because their sum is on the AQB spreadsheet in cell AA9, just to the right of the row for the coefficients. As it reminds you in the adjacent cell, the sum needs to be zero. If all is OK, then the results output appears automatically in the lower part of the 'RESULTS' box labelled 'Contrast'. They will not appear if the sum of the coefficients is not zero. The test statistic in this case is 't'.

For Prediction 1, that high motivation will increase performance, the first contrast is highly significant (since $t = 16.1$, d.f. = 216, $p < 0.001$): the mean values do indeed fall into the expected pattern because the contrast mean is positive (27.3). For the second contrast, we need to check whether it is independent of the first. We can do this by entering the second set of coefficients into the line below the first set (Fig. (ii)b). Once again, the box at the end of the row (cell AA10) will show zero if the two sets of coefficients are truly independent of one another (Fig. (ii)b). If they are, then we can transfer the second set of coefficients into the contrast row proper (Row 9, where the first set are) to obtain the result. We then find that, once again, the value of 't' is highly significant ($t = 8.0$, d.f. = 216, $p < 0.001$), so the mean values fall into the predicted pattern here too (again with a positive contrast mean).

Similarly for Predictions 2 and 3, we enter the coefficients and read off the result from the Contrast part of the Results box.

Because of the complexity of the predictions tested in this case, it is a good idea to set out the results in an ANOVA table, as in Fig. (iii); this contains all the details that are needed for readers to see exactly what the results are.

(a) Note that the width of some of the columns has been reduced to fit data and result into a single frame.

(b)

Figure (ii) (a) Results output and (b) checking the independence of contrast sets for a specific parametric two-way ANOVA using contrasts in AQB.

	ss	df	ms	F	prob
cells	1538.56	8	192.32	48.85	
Factor A (motivation)	1275.76	2	637.88	162.04	<0.001
contrast 1: high > (neutral + low)		1	1111.12	t = 16.8	<0.001
contrast 2: neutral > low		1	164.64	t = 6.5	<0.001
Factor B (temperature)	162.65	2	81.33	20.66	<0.001
contrast 1: cold > (warm + hot)		1	94.43	t = 4.9	<0.001
contrast 2: warm > hot		1	68.22	t = 4.2	<0.001
interaction (motivation x temperature)	100.14	4	25.04	6.36	<0.001
contrast 1: motivation difference, hot > (warm + cold)		1	105.47	t = 5.2	<0.001
contrast 2: motivation difference, warm > cold		1	0.01	t = -0.04	ns
within	850.32	216	3.94		
total	2388.88	224			

Figure (iii) Setting out the results from Fig. (ii) as a conventional ANOVA table for the purpose of a written report.

Testing for specific differences in a two-way classification using a parametric two-way ANOVA in R

We read in the data in the conventional format of a column for the data (`score`) and columns for each of the two factors (`motivation`, `temperature`). Make sure that R knows they are factors, and they have the correct order:

```
> motivation <- ordered(factor(motivation),levels=c("low","neutral","high"))
> temperature <- ordered(factor(temperature),levels=c("low","warm","high"))
```

Plot the mean values (Fig. (iv)), remembering to use 'beside=T' to put the bars next to one another rather than on top of one another:

```
> barplot(tapply(score,
   list(motivation,
   temperature),mean), beside=T)
```

There appears to be a strong pattern to the mean values. Now declare the contrasts you want to make, starting with the main effects (i.e. not the interactions):

```
> contrasts(motivation) <- cbind
   (c(-1,-1,2),c(-1,1,0))
> contrasts(temperature) <- cbind
   (c(2,-1,-1),c(0,1,-1))
```

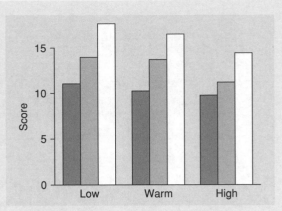

Figure (iv) Mean mathematical performance (score in a standard test) of students allocated to treatments in a two-way design of temperature (low, warm, high) and motivation (low, dark bars; neutral, grey bars; high, light bars).

Then run the ANOVA:

```
> m1 <- aov(score ~ motivation * temperature)
```

Check for normality and homogeneity – they seem fine:

```
> shapiro.test(resid(m1))
Shapiro-Wilk normality test
data: resid(m2)
W = 0.9939, p-value = 0.4987
> fligner.test(score~motivation*temperature)
Fligner-Killeen test of homogeneity of variances
data: score by motivation by temperature
Fligner-Killeen:med chi-squared = 3.087, df = 2, p-value = 0.2136

> summary(m1)
                       Df  Sum Sq Mean Sq  F value    Pr(>F)
motivation              2 1290.47  645.24 160.6455  < 2.2e-16 ***
temperature             2  235.47  117.74  29.3128 5.457e-12 ***
motivation:temperature  4   40.55   10.14   2.5242   0.04191 *
Residuals             216  867.57    4.02
```

The main effects of motivation and temperature are highly significant, and the interaction is just significant at the 0.05 level. Clearly there are patterns in each component: do they fall into the expected forms?

We ask for the summary of the linear model:

```
> summary.lm(m1)
Call:
aov(formula = score ~ motivation * temperature)
Coefficients:
              Estimate  Std. Error t value Pr(>|t|)
(Intercept)   13.17693     0.13361  98.624  < 2e-16 ***
motivation1    1.51707     0.09448  16.058  < 2e-16 ***
motivation2    1.30333     0.16364   7.965 9.37e-14 ***
```

```
temperature1                 0.53020    0.09448   5.612 6.09e-08 ***
temperature2                 0.85233    0.16364   5.209 4.43e-07 ***
motivation1:temperature1     0.10333    0.06680   1.547    0.1234
motivation2:temperature1     0.07893    0.11571   0.682    0.4959
motivation1:temperature2     0.10073    0.11571   0.871    0.3850
motivation2:temperature2     0.51020    0.20041   2.546    0.0116 *
--
Residual standard error: 2.004 on 216 degrees of freedom
Multiple R-squared: 0.6436,     Adjusted R-squared: 0.6304
F-statistic: 48.75 on 8 and 216 DF,  p-value: < 2.2e-16
```

The contrasts we asked for are the first four, all of which are in the correct direction (*t* values positive) and highly significant. Thus our predictions are supported by these data. The interactions that R reports are automatic, not the ones in which we are interested.

For the interactions we treat all the groups as if in a one-way ANOVA.

```
> treat <- factor(rep(c("lo.cold","lo.warm","lo.hot","nu.col",
  "nu.warm","nu.hot","hi.cold","hi.warm","hi.hot"), each=25))
> treat <- ordered(treat,levels=c("lo.cold","lo.warm","lo.hot",
  "nu.col","nu.warm","nu.hot","hi.cold","hi.warm","hi.hot"))
> contrasts(treat) <- cbind(c(-2,1,1,-2,1,1,2,-1,-1),c(0,-1,1,0,-1,1,0,1,-1))
> m2 <- aov(score ~ treat)
> summary.lm(m2)
Call:
aov(formula = score ~ treat)
Coefficients:
            Estimate Std. Error t value Pr(>|t|)
(Intercept) 13.17693    0.13361  98.624  < 2e-16 ***
treat1      -0.03896    0.09448  -0.412  0.68050
treat2      -0.14980    0.16364  -0.915  0.36098
treat3       0.47867    0.40083   1.194  0.23371
treat4       1.22937    0.40083   3.067  0.00244 **
treat5       0.09450    0.40083   0.236  0.81385
treat6       6.45622    0.40083  16.107  < 2e-16 ***
treat7       4.29312    0.40083  10.711  < 2e-16 ***
treat8       0.79679    0.40083   1.988  0.04809 *
--
Residual standard error: 2.004 on 216 degrees of freedom
Multiple R-squared: 0.6436,     Adjusted R-squared: 0.6304
F-statistic: 48.75 on 8 and 216 DF,  p-value: < 2.2e-16
```

We only asked for the first two, but R has filled in the rest automatically up to the limit of the degrees of freedom: we can ignore those. The two we asked for are in the wrong direction, and therefore not significant.

In conclusion, therefore:

- there is good evidence that higher temperatures reduce mathematical performance;
- there is good evidence that lowered motivation reduces mathematical performance;
- there is no evidence that higher temperature and lowered motivation combined cause reduced mathematical performance more than expected from their additive effects.

Box 3.9c   Medians (mean ranks) – a *general* non-parametric two-way ANOVA

General predictions in a two-way non-parametric ANOVA can be tested only if the sample sizes are equal for all groups. If they are not, then only specific predictions are testable.

Suppose a behavioural experiment measured the reaction times of different-sized insects of two kinds to a threatening stimulus. The factor *size* had three levels (small/medium/large), and the factor *taxa* had two levels (flies/beetles). Testing the residuals showed that they were far from normal (Shapiro–Wilk = 0.87, d.f. = 84, $p < 0.001$) and hence a non-parametric test is appropriate. The predictions were the usual general ones for a two-way design, involving the main effects (here, *size* and *taxa*) and their interaction:

- Prediction 1: beetles and flies will differ in reaction times,
- Prediction 2: large, medium and small insects will differ in reaction times,
- Prediction 3: there will be an interaction between size and insect type that affects reaction times.

Testing for general differences in a two-way classification using a non-parametric two-way ANOVA in AQB

In AQB, choose the 'np 2-way' worksheet, and then put the data in columns in the usual way (Fig. (i), and cf. Box 3.9b, Fig. (i)b). Note the 'labels' at the top that show how the two-way design (two sets of three groups) is arranged into two sets of three adjacent columns in the 'data' box below. Clicking the 'Go' button in the 'RESULTS' box will produce the output (if sample sizes are equal – otherwise a message-box reminder will appear). Like the non-parametric one-way

ANOVA in Box 3.5c, the analysis generates the test statistic H, but in this case we get one H-value for each of the main effects and one for the interaction. You can see that the medians (mean ranks) of neither the main effects nor the interaction differ significantly, so we cannot reject the null hypothesis for any of them.

Testing for general differences in a two-way classification using a non-parametric two-way ANOVA using R

There is no routine built into R for this test, and so we shall improvise. The data (`react`) are in one column, and the two factors (`insect`, `size`) in two other columns. Remember that you cannot do this test unless there are equal numbers of replicates in every box of the two-way design.

First we get a boxplot of the medians and ranges to visualize the data. For two-way data this is made easier by the lattice package (see Zuur *et al.*, 2009: chapter 8):

```
> library(lattice)
> bwplot(react ~ size|insect)
```

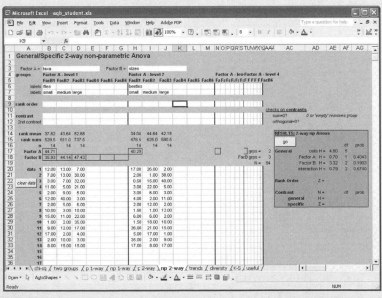

Note that the width of some of the columns has been reduced to fit data and result into a single frame.

Figure (i) Data entry and results output for a non-parametric two-way ANOVA in AQB.

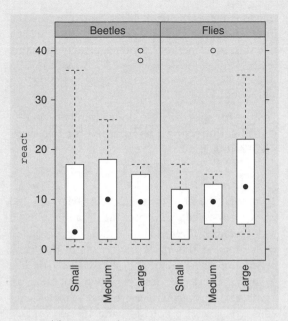

Figure (ii) Median (with interquartile ranges and extremes) reaction times (s) of beetles and flies of three different size categories

As Fig. (ii) shows, there seem to be some large differences, but also large variability in the measurements.

Now we need to make a one-way factor from all the combinations of the two-way factors (i.e. make the two insect levels and the three size levels into a factor with six levels representing the combinations). This is very easy in R using the ':' interaction operator:

```
> allgrps <- factor(insect:size)
```

The two-way analysis then consists of four steps:

- do a one-way Kruskal–Wallis test 'allgrps' combination factor
- do a one-way Kruskal–Wallis test on the row factor
- do a one-way Kruskal–Wallis test on the column factor
- subtract the H values in (b + c) from (a) to give the interaction H value

```
> kruskal.test(react,allgrps)
Kruskal-Wallis rank sum test
data: react and grps
Kruskal-Wallis chi-squared = 4.8034,
df = 5, p-value = 0.4403

> kruskal.test(react,insect)
Kruskal-Wallis rank sum test
data: react and insect
Kruskal-Wallis chi-squared = 0.6956,
df = 1, p-value = 0.4043

> kruskal.test(react,size)
Kruskal-Wallis rank sum test
data: react and size
Kruskal-Wallis chi-squared = 3.3186,
df = 2, p-value = 0.1903
```

Then the interaction H value is 4.8034 − (0.6956 + 3.3186) = 0.7892

Thus the ANOVA table is:

	H	d.f.	p
all groups	4.80	3	
rows (insects)	0.70	1	0.40
columns (sizes)	3.32	1	0.19
interaction	0.79	1	0.79

Thus we cannot reject the null hypothesis for any of our three hypotheses.

Box 3.9d   Medians (mean ranks) – a *specific* non-parametric two-way ANOVA

A team of researchers interested in the consequences of regular exercise for various measures of health and well-being investigated the joint effects of dietary fat and exercise (running on three or more days per week) on levels of blood cholesterol. They measured cholesterol in people who had low and high fat intake, and who did or did not run. They then carried out a non-parametric two-way ANOVA. The dependent variable being analysed is blood cholesterol level. There are two factors each with two levels: *diet* (high-fat/low-fat) and *exercise* treatment (no exercise/exercise).

The two-way design requires that each person being investigated has been allocated randomly to one diet treatment (low or high fat) and one exercise treatment (not running, or running). Testing the residuals for normality showed that they were normal (Shapiro–Wilk = 0.99, ns) but not homogeneous (Fligner $\chi^2 = 20.1$, d.f. = 120, $p < 0.001$), and hence a non-parametric approach is justified. Their predictions were:

- Prediction 1: running will reduce blood cholesterol,
- Prediction 2: a low-fat diet will reduce blood cholesterol,
- Prediction 3: the effect of running will be greater when on a low-fat diet.

We can test these predictions in two ways, either by predicting the *rank order* of the mean rank values, or by using single-degree-of-freedom *contrasts*. Since there are only two levels of each factor in the design, there is only one degree of freedom for each main effect, and one for the interaction. We shall see that, in a 2 × 2 design like this, both methods produce exactly the same result. Any other design (e.g. 3 × 2) will not do this.

Testing for specific differences in a two-way classification using a non-parametric two-way ANOVA in AQB

Testing rank order

To predict the rank order in AQB, select the 'np 2-way' worksheet, and enter the data in the usual way (*see* Box 3.9a). Then place the predicted rank order in the appropriate cells of the 'rank order' row (row 9) of the sheet (Fig. (i)). The example shows the test of the main effect of *exercise*. We expect cholesterol levels to be higher where no exercise is taken, and therefore place '2' in the two cells related to no exercise, and '1' in the two cells related to taking exercise.

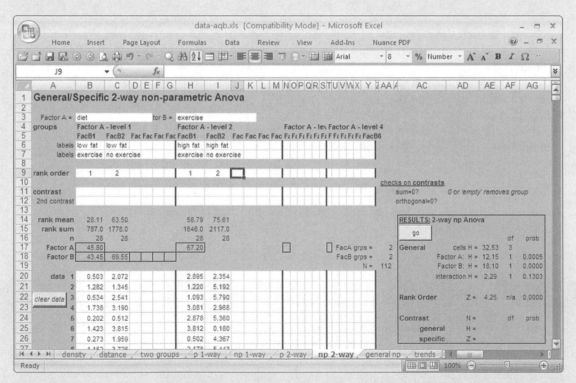

Figure (i) Data entry, setting up predicted rank orders and the resulting output for a specific two-way ANOVA in AQB.

Clicking 'Go' gives us the result. You can see that a general ANOVA is produced anyway, emphasising the fact that specific tests are a component of the general ANOVA. The rank order test has z as its test statistic: it is highly significant ($z = 4.25$, $p < 0.001$). A similar test for the impact of diet is also significant ($z = 3.49$, $p < 0.01$). In the case of the interaction, there is no way to make a rank-order prediction of all four cells, and therefore this prediction cannot be tested in this way.

Using contrasts
The second method involves setting up contrasts (*see* Boxes 3.5b and 3.9b) for the respective predictions as follows:

Exercise

Dietary fat		Yes	No
	Low	A	B
	High	C	D

Figure (ii) The two-way classification in the study.

Prediction 1: This can be cast in the form of an inequality, following the design of the study (Fig. (ii)), i.e.

$$B + D > A + C$$

and hence $(-1)A + (+1)B + (-1)C + (+1)D > 0$

The numbers in the brackets are the coefficients to use for the contrasts, to be placed in the 'contrast' row (Row 11) of the 'np 2-way' AQB sheet (Fig. (iii)). Clicking 'Go' produces the result, which is highly significant. Comparing it with the output in Fig. (i), we can see that it is identical.

Prediction 2: This can be described as:

$$C + D > A + B$$

and hence $(-1)A + (-1)B + (+1)C + (+1)D > 0$

Entering these coefficients produces the test result of $z = 3.49$, $p < 0.001$. Once again, this is

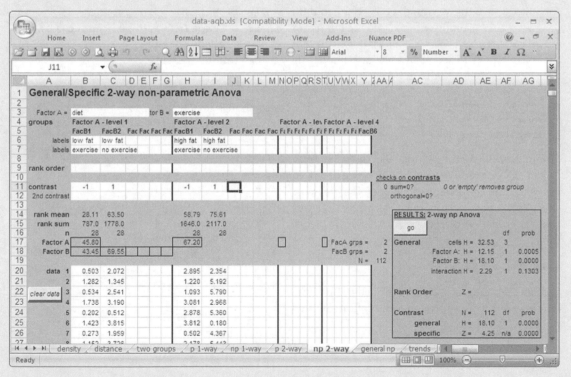

Figure (iii) Setting up the contrasts and resulting output for a specific two-way ANOVA in AQB.

identical to the previous result with a predicted rank order, because of the 2×2 design.

Prediction 3: This can be described in the form of differences:

$$B - A > D - C$$

and hence $\quad (-1)A + (+1)B + (+1)C + (-1)D > 0$

Entering these coefficients results in $z = -1.51$, $p = 0.065$. Thus there is no evidence from these data for the predicted interaction, although the p-value is quite close to significance: dietary fat and exercise appear to be purely additive in their effects. If we look at the general ANOVA in the 'RESULTS' box, we can see that no interaction of *any* kind seems to be present, so the lack of a significant outcome for our particular interaction is hardly surprising.

Testing for specific differences in a two-way classification using a non-parametric two-way ANOVA in R

As always, we have the data in one column (cholesterol), indexed by two columns of group codes (diet, exercise). First we can plot the mean values to see what they look like, using the library lattice to get the panel plots:

```
> library(lattice)
> bwplot(cholesterol ~
  diet|exercise)
```

The pattern looks interesting (Fig. (iv)).

To use a predicted rank order, we need to use the same coefficients developed for the AQB sheet (see above). Thus our first prediction was that exercise will reduce cholesterol levels, thus:

```
> c1 <- c(1,2,1,2)
```

We can then use the routine developed in Box 3.5d, but here we will add in the correction for tied ranks:

```
np.sp <- function(data,grps,coeff) {
r1 <- rank(data)
```

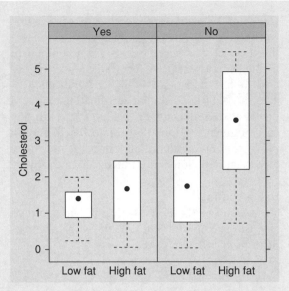

Figure (iv) Median (with interquartile ranges and extremes) blood cholesterol levels (arbitrary units) for runners ('yes') and non-runners ('no') given two different diets, a low-fat or a high-fat regime.

```
rsum <- tapply(r1,grps,sum)
nsum <- tapply(r1,grps,length)
L <- sum(coeff*rsum)
N <- length(data)
E <- (N+1)*sum(nsum*coeff)/2
V <- (N+1)*(N*sum(nsum*coeff^2) -
sum(nsum*coeff)^2)/12
Z <- (L-E)/sqrt(V)
r2 <- sort(data)
TC = 0
ii <- N-1
for (i in 1:ii) {
jj = i+1
tie = 1
for (j in jj:len) {
if (r2[i]==r2[j]) tie=tie+1 else j=len
}
TC = TC + ((tie^3)-tie)
}
TC = 1 - (TC/((len^3)-len))
Z = Z/sqrt(TC)
Z
}
```

We need to obtain a single factor with all the groups, to which the coefficients are applied. Thus:

```
> allgrps <- factor(diet:exercise)
> np.sp(cholesterol, allgrps, c1)
[1] 4.250945
```

For contrasts, we use the routine developed in Box 3.5d. Again we apply it to a set of all the groups, rather than the separate factors forming the two ways of the data.

This performs a Wilcoxon/Mann–Whitney test on the two groups with positive and negative contrasts, but is essentially the same test as that implemented in AQB. For example:

```
> c1<-c(-1,1,1,-1)
> npc(ch,allgrps,c1)
Wilcoxon rank sum test with
continuity correction
data: dneg and dpos
W = 1308, p-value = 0.06551
alternative hypothesis:
true location shift is less than 0
```

Although we have confined ourselves to a two-way analysis of variance here, the two-way model is only a particular case of a multifactor analysis of variance where there can be three, four or more levels of grouping. The principles underlying more complex analyses are the same, but as the number of levels of grouping increases it becomes more and more difficult to interpret the proliferating interaction terms. Moreover, the more levels of grouping that are included, the slimmer the chance they are all truly independent. Full discussion of these analyses can be found in Sokal & Rohlf (1995). In parametric analyses, as in the one-way case, it is also possible to incorporate covariates to control for factors on a constant interval scale.

3.3.4 Tests for a trend

As with analysis of differences, there are many tests that cater for trends. We shall introduce two simple ones here, both looking at the relationship between two sets of data. More complex versions of these tests allow multiple relationships to be tested at the same time, and we shall also look at how these work.

Correlation analysis

The first test is one of correlation. Correlation analyses calculate a test statistic known as a correlation coefficient, the two most commonly used being the parametric Pearson's product-moment correlation coefficient r, and the non-parametric Spearman rank correlation coefficient r_s (Box 3.10). A correlation coefficient quantifies the extent to which there is an association between two sets of data values. A large *positive* coefficient indicates a strong tendency for high values in one set to co-occur with high values in the other, and low values in one set to co-occur with low values in the other. A large *negative* coefficient indicates a strong tendency for high values in one set to co-occur with low

Box 3.10 Analysis of trends – tests of association (correlation)

An experimenter was interested in the association between the flight speed of house skylarks (*Alauda arvensis*) and their body sizes. To see whether size and speed were associated, she took a number of individual birds for which she already had some body weight data and measured their maximum flying speed in a flight chamber. Two points are important about these measures. The first is that the experimenter had not pre-determined the body weights in any way (she hadn't deliberately chosen birds of a particular weight, or fed any up to make them heavier in an experimental manipulation; she had simply caught some birds arbitrarily with respect to their weight). The second is that, should an association between weight and flight speed emerge, the cause-and-effect relationship between the two could conceivably be in either direction: heavier birds may be stronger and so be able to fly faster (or have to spend more energy getting up speed, so fly more slowly), or birds that fly faster may burn up fat reserves and so be lighter. In other words, weight might affect flight speed, or flight speed might affect weight, or both could depend on some other (unmeasured) variable: we can't tell a priori. For these reasons, the appropriate trend analysis here is a correlation. Depending on the distribution of the data (parametric correlation requires both variables to be normally distributed, and for both together to show a bivariate-normal distribution – a three-dimensional (3D) bell shape), we could perform either a parametric or a non-parametric correlation.

This analysis is the only case where testing the raw data for normality is correct. Body *mass* seems to be

normally distributed (Shapiro–Wilk = 0.98, d.f. = 99, ns), but flight *speed* is not (Shapiro–Wilk = 0.95, d.f. = 99, $p < 0.001$). We cannot test for bivariate-normal distributions without extremely large sample sizes. We shall therefore do both a parametric and a non-parametric correlation with the same data to demonstrate their similarity.

Let's suppose we have good reason for thinking heavier birds will fly faster because they are more powerful. We thus predict a positive correlation between weight and flight speed (H_1) (the null hypothesis, H_0, is that there is not a positive correlation). This is an a priori prediction, so a specific test is appropriate.

Correlation analysis in AQB

In AQB, select the 'trends' worksheet, and enter the weight and speed data in two columns of paired values, as in Fig. (i). The 'RESULTS' box contains output for both regression and correlation analyses, as in many statistical packages, but these are different analyses with different

Figure (i) Data entry and results output for trend analyses in AQB.

underlying assumptions, so, for the present case, ignore the regression output. The 'normal' correlation (called a Pearson product-moment correlation in full) appears automatically. Since we are testing the specific prediction of a positive trend, we use the 'specific' rather than 'general' output, which gives us the one-tailed probability associated with the outcome *if* the value for the test statistic, *r*, has the predicted sign (positive for a positive trend, negative for a negative trend). It clearly has here, and hence the result is highly significant here.

For a non-parametric (Spearman rank) correlation, click the 'Rank' button. With large ($n > 30$) sample sizes, as here, the difference in outcome between Pearson and Spearman correlations is negligible.

Correlation analysis in R

Read in the data as two columns, `mass` and `speed`. We will first plot the data:

```
> plot(mass, speed)
```

It certainly looks as if there is a clear relationship (Fig. (ii)), but as we have seen in the text, appear-

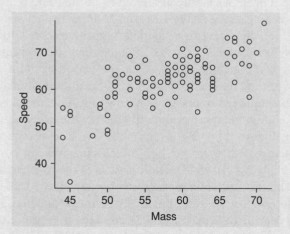

Figure (ii) Flight speed (m s⁻¹) and body mass (g) of skylarks (*Alauda arvensis*)

ances can be deceptive. To do the default Pearson correlation test, simply type:

```
> cor.test(mass,speed)
```

Here we are making the prediction of a positive relationship, and so we have to ask for a one-tailed test:

```
> cor.test(mass,speed,
  alternative=c("greater"))
```

```
Pearson's product-moment correlation
data: mass and speed
t = 9.1899, df = 97, p-value =
3.775e-15
alternative hypothesis:
true correlation is greater than 0
95 percent confidence interval:
0.581939 1.000000
sample estimates:
     cor
0.6822262
```

For a non-parametric correlation, the Spearman correlation, we use the same test but switch it to the Spearman:

```
> cor.test(mass,speed,
  method=c("spearman"),
  alternative=c("greater"))
```

```
Spearman's rank correlation rho
data: mass and speed
S = 62095.73, p-value = 5.757e-12
alternative hypothesis: true rho is
greater than 0
sample estimates:
     rho
0.6159819
Warning message:In
cor.test.default(mass, speed, method
= c("spearman")) : Cannot compute
exact p-values with ties
```

Clearly, the conclusion from these tests is that there is indeed strong evidence of a positive correlation between these two variables.

values in the other and vice versa. Correlation coefficients take a value between +1.0 and −1.0, with values of +1 and −1 indicating, respectively, a perfect positive or negative association. 'Perfect association' means that every value in one set is predicted perfectly by values in the other set. A coefficient of 0 indicates there is no association between the two sets of values so that values in one set cannot be predicted by those in the other. If a correlation is significant, it implies that the size of the coefficient differs significantly (positively or negatively) from zero, the value expected under the null hypothesis.

As a parametric test, the Pearson product-moment correlation is subject to a number of assumptions about the data being analysed. The two sets of data must be normally distributed individually *and* jointly (a bivariate normal distribution in the jargon), both must be measured on a constant interval scale, and the relationship between them must be linear. Satisfying these assumptions can be a tall order and the Pearson correlation is probably used more liberally than it should be. Being a non-parametric test, the Spearman rank correlation can be used with ordinal (ranking) or constant interval measurements and, of course, is not sensitive to departures from normality. Importantly, the relationship also need not be linear, but merely continuously increasing or decreasing (monotonic); thus r_s will test for many sorts of curved patterns of association.

While we can test for trends with correlation analyses, we must interpret them with care. Two things in particular should always be borne in mind. First, a correlation does not imply cause and effect. While we may have reasons for *supposing* that one measure influences the other rather than vice versa (*see* earlier discussion of x and y measures in trend analysis), a significant correlation cannot be used to confirm this. All it can do is demonstrate that two measures are associated. A well-known example illustrates the point. Suppose we acquired some data on the number of pairs of storks breeding in Denmark each year since 1900, and also the number of children born per family in Denmark in the same years. Plotting a scattergram, with breeding storks as the x-axis and babies as the y-axis, and calculating a correlation coefficient reveals a significant positive correlation at $p < 0.001$ between the two measures. Do we conclude that storks bring babies? Of course not! All we can conclude is that, over the period examined, there is some association between the number of breeding storks and the human birth rate; perhaps both species simply reproduce more during long, hot summers! Second, as we have said, correlation analyses assume that associations between measures are linear (or reasonably so) if the test is parametric, or at least monotonic if the test is non-parametric. If they are neither, a lack of significant correlation cannot be taken to imply a lack of association. This is made clear in Fig. 3.4 in which the relationship between two measures is U-shaped. A correlation coefficient for this would be close to zero, but this doesn't mean there is no association. The book by Martin & Bateson (1993) contains a very useful discussion of these and other problems concerning correlation analyses.

Linear regression

Correlation analyses allow us to judge whether two measures are associated, but that's all. Two important things that they do not allow us to establish are: (a) whether changes in the value of one measure *cause* changes in the value of the other and (b) anything quantitative about the association.

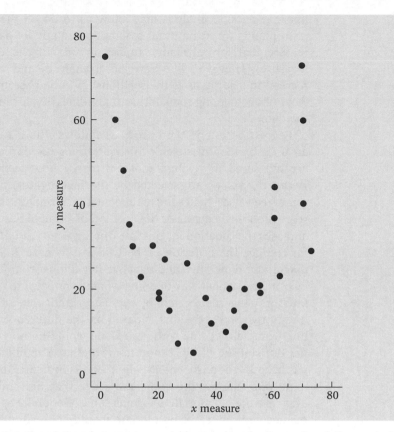

Figure 3.4 Associations between two variables need not be linear. Correlation analysis would not reveal a significant trend, but an association between the variables appears to exist (*see text*).

Cause and effect. In the case of the storks and babies above, it is fairly clear that changes in x (the number of breeding storks) do not cause changes in y (the number of human babies): the association arises either through some indirect cause-and-effect relationship (probably that habitat destruction for storks and the demographic transition to smaller family sizes in humans both happened over the course of the twentieth century) or through trivial coincidence. In many cases, though, it is not so clear whether there is or is not a direct cause-and-effect relationship between x and y. The best way to decide is to do an experiment where, instead of merely measuring x and y as in correlation analysis, we experimentally change x-values and measure what subsequently happens to y. If we see that y changes when x is changed, we can be reasonably happy with a cause-and-effect interpretation. This is quite different from correlation analysis.

Quantitative relationships. In addition, correlation analysis does not allow us to say much about the quantitative relationship between x and y (if x changes by n units, by how much does y change?) or allow us to predict values not included in the original trend analysis (e.g. can we predict the response of an insect pest to a 40 per cent concentration of pesticide when an analysis of the effect of pesticide concentration goes up to only 30 per cent?). With certain qualifications, linear regression analysis may allow us to do all the above. The qualifications arise mainly from the usual assumptions made by parametric tests, though the requirement for normality applies to one of the data sets only (the y-axis measure). However, like correlation analysis, and as its name implies, linear regression also requires the relationship to be passably linear, though there are ways of overcoming some forms of non-linearity, for instance by log-transforming the data.

Regression analysis proceeds as follows: the x-values are decided upon in advance by the investigator to cover a range of particular interest, and y-values are measured in relation to them to see how well they are predicted by x. Because x-values are selected by the investigator, they are regarded as fixed, error-free values: hence the requirement of normality only on the y measure. Linear regression then calculates the position of a line of best fit through the data points and uses the equation for this line (the *regression equation*) to predict other values for testing. The criterion of 'best fit' in this case is the line that minimises the magnitude of positive and negative deviations from it in the data – a more precise, mathematical way of doing what we attempt to do when drawing a straight line through a scattergram by eye. The significance of a trend can be assessed in one of two ways: the first is based on the difference in the *slope* of the best fit line from zero (a line with zero slope would be horizontal) and is indicated by the test statistic t; the second tests whether a significant amount of variation in y is accounted for by changes in x, and is indicated by the test statistic F. There is no point in quoting both tests, since they are really the same thing: $F = t^2$! They are merely alterative formulations. Generally speaking, if the trend fails to reach significance, the best fit line should not be drawn through the points of the scattergram. Box 3.11 shows how to perform a linear regression in AQB and R.

Multiple regression. So far we have just dealt with simple linear regression involving a single x and single y variable. It is not difficult to imagine, however, that we might have several candidate independent (x) variables that could potentially explain variation in our dependent (y) variable. If so, we need some way of taking them into account simultaneously so we can control for the effects of other variables when seeking effects of any particular one. Happily, there is a relatively straightforward set of techniques for doing this, which come under the umbrella term *multiple regression*. As ever, there are various "dos" and "don'ts" associated with the procedures, but Box 3.12 outlines a reasonably robust way of setting about the business.

Data reduction

We conclude our summary of statistical procedures on a somewhat different note from the preceding significance tests. As we have seen in the case of analysis of variance and regression, we are often confronted with several variables that could individually, or in various combinations and interactions, explain variation in a dependent variable. Multifactor and multivariate analysis of variance and multiple regression are one set of procedures that allows us to deal with this kind of situation. However, there is another way, which is not itself a significance test of any kind, but is instead a statistical technique for

Box 3.11   Analysis of trends – linear regression

A microbiologist was interested in the effect of the concentration of a particular vitamin on the growth of the fungus *Aspergillus*. He therefore did an experiment in which the concentration of vitamin solution added to the fungal culture was varied and subsequent growth of the culture measured (as mg dry weight of fungus).

As the vitamin, and its requirement by the fungus, is known, prior information suggests that there should be a relationship between concentration and growth and that it should be positive, with higher concentrations leading to greater growth. We thus have a specific prediction of a positive relationship between the two variables (H_1) and a null hypothesis (H_0) of no positive relationship. Unlike the correlation analysis in Box 3.9a, however, we are here predicting a very definite cause-and-effect relationship, because we have *experimentally fixed (manipulated)* vitamin concentrations in order to see the impact of these manipulations on growth. In statistical terminology, therefore, vitamin concentration is the independent variable (the predictor), and growth the dependent variable (the response). Analysis of the residuals (*see* Box 3.1b) shows that they can be regarded as normally distributed (Shapiro–Wilk = 0.98, d.f. = 25, ns).

Performing regression analysis in AQB

On the 'trends' worksheet in AQB, data for regression are entered in paired columns exactly as for correlation (Fig. (i)). The output appears automatically in the 'RESULTS' box. The slope (b) of the regression line and its standard error (here, $b = 0.184 \pm 0.014$) is given at the top, along with the equation describing the line (in which b is the coefficient determining the slope, and Y is the *growth* value predicted by the equation for any given vitamin concentration). Next the value of r^2 is given, the proportion of the variation in growth accounted for by the regression relationship with vitamin concentration. There are then two ways of assessing the significance of the regression: by an ANOVA (*see* Boxes 3.9a–d), which calculates an F-ratio (here, $F = 170.5$, d.f. = 1,23, $p < 0.0001$), or by a t-test (which is itself a form of ANOVA [t is just the square root of F] – see text for the difference in what they test). Here, $t = 13.1$, d.f. = 23, $p < 0.0001$. An important point is that the ANOVA version with its F-test is a general test only, whereas the t-test can be general (two-tailed) or specific (one-tailed), depending on the nature of the prediction.

There is a further possibility in AQB. We can use the regression to make a prediction about new values of Y (here *growth*) for any given new value of X (*vitamin*). The new value of X

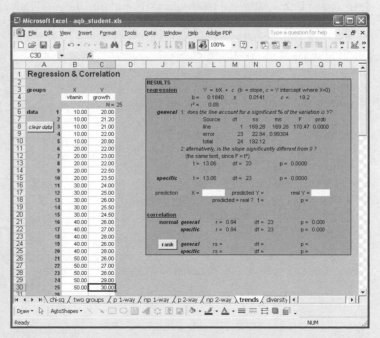

Figure (i) Data entry and results output for linear regression analysis in AQB.

must lie within the range of X-values used to generate the regression equation, otherwise the prediction can be very inaccurate indeed. This risk of inaccuracy arises from two potential sources of error: first, there is some uncertainty in the calculated value of the slope – indicated by its standard error – which will magnify the uncertainty of the Y-value the further away we get from the middle of the line (the mean values of X and Y); second, the prediction error increases towards the ends of the regression line simply because there are usually fewer data points out there compared with the middle of the line.

Bearing this caveat in mind, we can enter a new X-value into the 'X=' box on the 'prediction'

Figure (ii) Predicting and testing new values of Y for given new values of X in regression analysis using AQB.

line of the AQB screen – here we've entered a new value of 45 (Fig. (ii)) – and the new value of Y predicted by it will appear as 'predicted Y' next to it. The experimenter can then carry out some new tests to see if this is upheld by the data. By entering the newly measured Y from his test into the 'real Y=' box, the experimenter can perform a *t*-test to see if it differs significantly from the value predicted by the regression equation. The *t*-value and its associated probability appear next to 'predicted = real?' on the next line in Fig. (ii). Here we can see that the observed value of 28 is not significantly different from the predicted one of 27.5; so the regression equation seems to be a good predictor of new Y-values, at least for X-values within its range.

Performing regression analysis in R

The data for regression analysis are in two columns, forming *X–Y* pairs, `vitamin` and `growth`. Import into R in the usual way (Box 2.2). We will plot the data as a scatterplot first, and then add the dashed line of best fit from the linear model (`lm`):

```
> plot(vitamin, growth)
> abline(lm(growth ~ vitamin),lty=2)
```

The result (Fig. (iii)) looks pretty convincing. Then we carry out the regression. Because we have not declared either variable as a factor, and both are numeric, R knows that both are continuous constant-interval measurements.

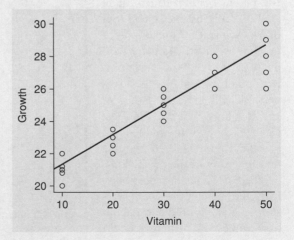

Figure (iii) Growth (μg day^{-1}) of *Aspergillus* fungus on a diet supplemented with different levels of vitamin B$_2$ (mg l^{-1} agar).

```
> m1 <- lm(growth ~ vitamin)
```

Note the standard format of this linear model ('lm') where the response (on the left-hand side) is predicted by ('~') the predictor variable of the right-hand side. Now we need the details:

```
> summary(m1)
Call:
lm(formula = growth ~ vitamin)
Coefficients:
            Estimate Std. Error t value Pr(>|t|)
(Intercept) 19.20000    0.46741   41.08  < 2e-16 ***
vitamin      0.18400    0.01409   13.06 4.03e-12 ***

Residual standard error: 0.9965 on 23 degrees of freedom
Multiple R-squared: 0.8811,     Adjusted R-squared: 0.8759
F-statistic: 170.5 on 1 and 23 DF, p-value: 4.031e-12
```

The estimate for the 'intercept' here really is the intercept in the model $y = mx + c$. The t-value tests whether it is significantly different from zero, and it clearly is. The estimate for 'vitamin' is the estimated slope of the effect of the predictor, vitamin, and the t-value tests whether it is significantly different from zero. The Multiple R-squared is the fraction of the variance in the response variable (growth) explained by the model (here, just by the single predictor, vitamin). The 'adjusted R-squared' is a similar calculation but based on variances rather than sums-of-squares: it is better to use this value, since it is 'unbiased', in the statistical jargon.

The conclusion from both packages, then, is that increasing vitamin concentration does indeed increase the growth of *Aspergillus*, and the regression equation quantifies the amount by which growth is increased by any given increase in vitamin concentration, within the limits of the experiment.

Box 3.12   Multiple regression

Often we are interested in the predictability of some measure from a number of other variables. For example, suppose we know that the body size of, and the number of eggs laid by, an insect (*Cynips divisa*) that lives inside a plant gall on oak leaves (*Quercus robur*) are very closely related to the size of gall in which it develops. The galls grow on the veins underneath the leaves. What determines gall size? We suspect that the availability of resources from the plant might be one influence, and also the extent of competition for those resources (how many galls are there per leaf or per vein?). Thus we obtain a data set taken from a large number of leaves, where we measure the width of the gall (*gdiam*) as a measure of gall size, the length of the leaf (*llen*), the width of the leaf (*lwid*), the number of galls on the leaf (*ngalls*), the number of galls on a vein (*vgno*), the *order* in which each gall developed on a vein (first gall, second, third, etc.) and the distance of each gall from the midrib of the leaf (*vdist*). Thus we might predict a priori that gall diameter (*gdiam*) should:

■ increase with leaf length (*llen*) and/or leaf width (*lwid*), because bigger leaves will offer more resources;
■ decrease with the extent of competition either on the whole leaf (*ngalls*) or on an individual vein (*vgno*);
■ decrease with increasing *order* on a vein and, for a given order, decrease with distance from the midrib (*vdist*) (presuming that the resources for growth come down the midrib and along the veins – something that has been demonstrated in other gall systems).

Here we have one response variable (gall diameter, *gdiam*), and a set of independent predictor variables (*llen*, *lwid*, *ngalls*, *order*, *vngo*, *vdist*). The predictions can only be tested observationally since no manipulations are possible in this system. Even though regression is a technique developed for situations where the experimenter determines the *x*-values, it is very common for it to be used in other situations too, and we shall do this here. We shall check for the normality of the residuals when we have settled on the best model for predicting the dependent variable. To find the best model, we shall use an approach known as *backwards deletion*, a procedure whereby the different available predictors are deleted (or not, depending on their individual contribution to the model) successively to the analysis according to certain statistical rules. In doing so, we will encounter some important principles of modern statistical analysis. We will discuss them separately (Box 3.15) because they are too important to bury within a discussion of multiple regression. As you become more familiar with statistical analysis, you will realise that everything is becoming simpler as the statistical toolbox becomes more and more comprehensive. Analysis using the Generalised Linear Models of R is an important part of this process.

The principle needed here is parsimony, the idea that science seeks the simplest explanation (see Box 3.15). We will delete predictors from our multiple regression if this reduces the AIC of the subsequent model, implemented by the R command 'step'.

We will try to circumvent the problem of order effects (see Box 3.15) by using biological knowledge to help in dictating the order of the predictors in the model. Here we use scale, starting with leaf-level predictors (leaf length, leaf width, gall density), and then vein-level predictors (number of galls on the vein, distance of a gall along a vein, order of the gall on the vein).

Model simplification in R

Import the data into R in the usual way, with each variable in a separate column.

```
> d1<-read.table(file.choose(),header=T)
> attach(d1)
> names(d1)
[1] "llen" "lwid" "ngalls" "vgno" "vdist" "order" "Gdiam"
```

Now we fit the maximal model. Each of these predictors is a constant-interval measurement, either on a continuous scale, or on an integer scale. Thus none are factors: this is a multiple regression.

```
> m1 <- lm(Gdiam ~ llen + lwid + ngalls + vgno + vdist + order)
> summary(m1)
Call:
lm(formula = Gdiam ~ llen + lwid + ngalls + vgno + vdist + order)
Coefficients:
            Estimate Std. Error t value Pr(>|t|)
(Intercept)  3.9028865  0.0936013  41.697  < 2e-16 ***
llen         0.0085006  0.0013583   6.258 4.26e-10 ***
lwid        -0.0049404  0.0020429  -2.418  0.01563 *
ngalls      -0.0405352  0.0045150  -8.978  < 2e-16 ***
vgno         0.0835889  0.0311068   2.687  0.00723 **
vdist        0.0023154  0.0008319   2.783  0.00540 **
order       -0.3144411  0.0407203  -7.722 1.41e-14 ***
---
Residual standard error: 1.284 on 4434 degrees of freedom
```

```
 (4648 observations deleted due to missingness)
Multiple R-squared: 0.05336,    Adjusted R-squared: 0.05208
F-statistic: 41.65 on 6 and 4434 DF,  p-value: < 2.2e-16
```

All predictors are significant, but the model accounts for only just over 5 per cent of the variation in the data. Like a lot of ecological data sets, there are plenty of sources of variation that were not measured. Since there are lots of observations, however, the effects are still apparent.

We can obtain the AIC of this maximal model by asking:

```
> AIC(m1)
[1] 14831.08
```

Now we use the 'step' command, that tries to delete predictors based on the AIC:

```
> step(m1)
Start: AIC=14831.08
Gdiam ~ llen + lwid + ngalls + vgno + vdist + order
         Df Deviance    AIC
<none>         7308.1  14831
- lwid    1    7317.7  14835
- vgno    1    7320.0  14836
- vdist   1    7320.9  14837
- llen    1    7372.6  14868
- order   1    7406.4  14888
- ngalls  1    7440.9  14909

Call: glm(formula=Gdiam~llen+lwid+ngalls+vgno+vdist+order)
Coefficients:
(Intercept)   llen       lwid      ngalls    vgno      vdist      order
   3.902887   0.008501  -0.004940  -0.040535 0.083589 0.002315 -0.314441
Degrees of Freedom: 4440 Total (i.e. Null);  4434 Residual
  (4648 observations deleted due to missingness)
Null Deviance:       7720
Residual Deviance: 7308          AIC: 14830
```

It turns out that the full model is the minimal sufficient model: the step procedure could not find any predictor to delete that would decrease the AIC.

The coefficients are the most important part of the output, because they show how the predictor affects the response variable. The intercept is the overall mean gall diameter. There are positive effects of leaf length (llen, with a slope of 0.0085 ± 0.0013), the number of galls per vein (vgno, slope 0.084) and distance along the vein (vdist, slope 0.0023); and negative effects of leaf width (lwid, slope −0.0049), number of galls per leaf (ngalls, slope −0.041) and order on the vein (order, slope −0.314). The largest slope is that of gall order on the vein, the largest effect on gall diameter. This is interpretable in terms of competition for nutrients.

Since we have the coefficients and the t-values, we can test our a priori predictions about the sign of the effects of these predictors. Clearly, as we predicted, there is evidence of the impact of both the amount of resource (leaf length) and resource competition (number of galls, order), but some coefficients are not in the predicted direction (leaf width, number of galls on a vein). This might stimulate new hypotheses to be tested with new data.

collapsing several interrelated variables down to one or two composite variables that can then be treated as new data (either as response (dependent) or predictor (independent) variables). With the caveat that resulting composite variables can sometimes be complicated to interpret, this can be an extremely useful way of rendering multivariate data manageable for analysis. By way of example, we summarise a commonly used such procedure, known as *principal components analysis*, in Box 3.13.

Box 3.13	Principal components analysis – creating a reduced set of variables from a larger set of intercorrelated ones

A common problem in many studies is that data are collected for a large number of variables that are interrelated, so looking for differences or trends in each of them separately would result in a high degree of non-independence between tests. However, there are statistical techniques that help overcome this by producing single variables that are composites of the original separate ones. One of these techniques is known as principal components analysis (PCA).

Unlike the tests we've encountered so far, PCA isn't an hypothesis-testing method at all, but a method for reducing the number of variables to something more manageable (the so-called principal components). There is no response variable in this situation. Thus, for us, its main attraction is that we can replace many variables with just two or three that (a) capture the main features of the data, and (b) are (by definition) statistically independent of one another. These new variables can then be used like any others in statistical tests of hypotheses, usually representing an enormous gain in clarity and efficiency. It is best learned through an example, because mathematically it is complicated!

Suppose a biologist was interested in analysing the differences in morphology among the species of flies that he worked on, and therefore he measured a set of 17 variables on a large number of individuals of many species. He then took the averages of each of the variables for each species, but kept males and females separate. This resulted in the data set of Fig. (i). Details of the variables are in the figure legend.

Doing principal components analysis in R

We read the data we want to use into a dataframe (d1); they consist of columns of variables in the usual way. Then the analysis is done using the command:

```
> m1<-prcomp(d1, scale=T)
> summary(m1)
Importance of components:
                       PC1    PC2    PC3    PC4
Standard deviation     3.566  1.116  0.6170 0.4832
Proportion of Variance 0.848  0.083  0.0254 0.0156
Cumulative Proportion  0.848  0.931  0.9562 0.9717
```

The columns labelled 'PC1', 'PC2', etc. are new independent (uncorrelated) axes of variation, in order of their importance in accounting for the data. We have omitted PC5 to PC15 because, as is clear from the output, the first four already account for more than 97 per cent of the variation in the data. The first one in particular contains almost 85 per cent of the variation, and the second just over 8 per cent.

We can plot these first two axes in a special kind of plot called a biplot (Fig. (ii)). By default each point is plotted as a row number from the data matrix; we have suppressed these in favour of simple dots here because otherwise the whole plot is obscured.

```
> biplot(m1,xlabs=rep(c("."),each=452))
```

SEX	WL	WW	HW	THW	THL	THH	THVO	T2	T3	T4	FU	LABR	PR	LL	PS
F	6.61	1.98	2.19	1.52	2.55	1.98	7.74	1.83	1.79	1.7	1.14	0.89	0.64	0.52	19.3
F	6.75	2.24	2.22	1.59	2.65	2.02	8.51	1.92	1.88	1.7	0.88	0.78	0.48	0.62	28
F	7.37	2.55	3.07	2.29	3.35	2.56	19.6	2.68	2.67	2.48	1.34	1.24	0.85	0.86	28
M	7.4	2.34	3.05	2.38	3.44	2.72	22.3	2.68	2.45	1.98	1.41	1.26	0.88	0.84	26
F	7.64	2.56	3.14	2.38	3.54	2.71	22.8	2.93	2.93	2.62	1.52	1.44	0.96	0.95	29
M	6.51	2.22	2.73	2.14	3.13	2.43	17	2.41	2.12	1.76	1.31	1.31	0.92	0.79	27.3
M	10.7	3.35	4.1	3.55	5.22	4.25	79	4.92	4.71	3.9	1.63	1.8	1.19	1.26	44
F	8.6	3.23	4.28	3.05	4.39	3.2	42.9	4.04	3.91	3.45	1.84	1.73	1.16	0.87	23
F	8.22	3.06	4.17	2.99	4.32	3.28	42.7	4.07	3.94	3.44	1.74	1.63	1.2	0.8	24.7
M	7.82	2.84	4.05	2.94	4.2	3.15	38.9	3.55	3.44	2.85	1.73	1.64	1.13	0.81	26
F	7.39	2.04	1.75	1.12	2.05	1.66	3.85	0.49	0.98	1.27	0.64	0.43	0.35	0.54	26
M	6.4	1.76	1.7	1.16	2.2	1.92	5.36	0.55	0.5	0.78	0.59	0.39	0.34	0.55	26
F	8.65	2.84	3.08	2.38	3.88	2.76	25.5	3.2	3.25	3.05	1.09	0.76	0.57	1.17	54
M	8.72	2.93	3.46	2.98	4.39	3.46	45.3	4.51	4.33	3.83	1.28	1.31	0.81	1.19	40
F	9.92	3.65	3.58	2.9	4.54	3.49	45.9	4.62	4.65	4.09	1.46	1.39	0.9	1.32	39.5

Figure (i) A data set of the morphology of a set of species of flies. The rows are labelled with the genus name only (not shown), followed by M (males), or F (females). The variables were: wing length (WL), wing width (WW), head width (HW), thorax width (THW), thorax length (THL), thorax height (THH), thorax volume (THVOL = THW*THL*THH), the widths of abdominal segments 2–4 (T2, T3, T4), the lengths of the parts of the proboscis (FU, LABR, PR), width of the fleshy pad at the end of the proboscis (LL), and the number of sucking channels on the pad (PS).

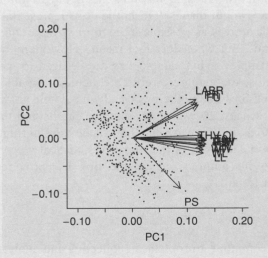

Figure (ii) Principal components biplot of the first two axes of a principal components analysis of morphological variables of a set of species of flies. The plotted points represent the scores along the first two axes of each individual species. The vectors represent the loadings (eigenvectors) of the original variables along the first two axes.

We can see the original variables plotted as vectors (the arrows). This allows us to interpret what the axes represent. We see that all arrows point to the right and are therefore positively correlated with the x-axis (PC1); most of them are pointing directly along the x-axis, except for variables associated with the proboscis (see legend to Fig. (i)), which are at an acute angle to the x-axis. Thus PC1 represents body size, an increase in all measurements. If we project the arrows along the y-axis, only the proboscis variables produce any length of arrow: hence PC2 represents the proboscis shape. By definition in PCA, axes are uncorrelated with one another. Thus PC2 represents variation in the proboscis not connected with body size – it is the shape axis. The only arrow pointing downwards is PS, the number of feeding channels in the fleshy pad at the end of the proboscis (hence negatively correlated with PC2): all other proboscis variables measure the lengths of its parts, and point upwards (positively correlated with PC2). Thus species that plot high up on the graph have long tongues with few feeding channels, while those at the bottom have short tongues with lots of feeding channels.

Thus 15 variables here can be replaced by just two which represent body size and proboscis shape, and which account for 93 per cent of the variation in the data. By doing so, we replace 15 intercorrelated variables with two uncorrelated independent variables that can be used separately with confidence to test hypotheses in statistical tests.

We extract PC1 and PC2 scores as variables for use in tests by typing:

```
> PC1 <- predict(m1)[,1]
> PC2 <- predict(m1)[,2]
```

3.4   Testing hypotheses

The previous section has introduced an armoury of basic significance tests with which we can undertake confirmatory analyses of differences and trends. Knowing that such tests are available, however, is not much use unless we know how and when to employ them and gear our data collection to meet their requirements. It is important to stress again, therefore, that the desired test(s) should be borne in mind from the outset when experiments and observations are being designed and the data to be collected decided upon.

3.4.1   Deciding what to do

Having arrived at some predictions from our hypotheses, we must decide how best to test them. This sounds straightforward in principle but involves making a lot of careful decisions. Are we looking for a difference or a trend? What are we going to measure? How are we going to measure it? How many replicates do we need? What do we need to control for? There is no general solution to any of these problems; the right decision depends entirely on the prediction in hand and the material available to test it. In a moment, we shall go back to the predictions we derived from our observational notes to see how we can test some of them. Before doing that, however, we should be aware of some important principles and pitfalls of experimental/observational design and analysis.

Significance, sample sizes and statistical power

As should be obvious from what we've said already, much hinges on the quality of the data sample we have at our disposal. It should be as representative of the population from which it derives as possible if we're to stand a chance of coming to sensible conclusions. But what does that mean? How many data values make a representative sample? What sample size do we need to make our tests for differences or trends reasonably powerful, i.e. actually capable of detecting the effects we're testing for? Sadly, there is usually no simple answer. Vague rules of thumb, such as 'at least 20 values per sample', can be and have been suggested (e.g. Dytham, 2010), but they are just that, vague rules of thumb.

What is really needed is a formal analysis of the so-called statistical *power* of a significance test in relation to the sample size available. Box 3.14 provides a summary introduction to power tests.

Some dangerous traps

Confounding effects. One of the commonest problems in collecting and analysing data is avoiding so-called confounding effects. Confounding effects arise when a factor of interest in an investigation is closely correlated with some other factor that may not be of interest. If such a correlation is not controlled for, either in the initial design of an investigation or by using suitable techniques during analysis, such as analysis of covariance (*see* Box 3.6), the results will inevitably be equivocal and any potential conclusions compromised. For example, suppose

Box 3.14   Power tests

When we carry out an experiment, we are trying to discern a pattern (a difference or a trend) in the face of variation from a number of sources. We use a statistical test to guide us, which in essence boils down all the data into a single number, the test statistic for that analysis, the distribution of which is known *under the null hypothesis*. Remember that the *p*-value we get from the test statistic is the probability of obtaining the data we got *if the null hypothesis is true*.

We can think about statistical testing in terms of discriminating a signal from the noise of natural variability. This noise can cause us to make errors in the conclusions of our experiment, since we don't know what the true situation is, i.e. whether the null hypothesis is true or not. This sets up four possibilities (Fig. (i)).

We can make two correct decisions (reject the null hypothesis correctly because it is actually false, or not reject the null hypothesis correctly because it is in fact true). We can also make two mistakes. One (a Type I error) is important because we are drawing a conclusion – we reject (erroneously) the null hypothesis in favour of its alternative, H_1. All statistical tests of hypotheses are designed to fix the probability (denoted α) with which such mistakes are made – it's the familiar threshold for significance, usually 0.05. We're prepared to be wrong 5 per cent of the time, but that's low enough to be reasonable.

The other mistake (a Type II error) is when we erroneously fail to reject the null hypothesis even though it is false. This is less serious since we are not drawing any conclusions about the data (unless we would like to make the much stronger claim that the null hypothesis is *actually correct*, an error of logic in scientific procedure). The rate at which this mistake happens, β, is a function of a number of factors, including α, the design of the experiment, the test used, the sample sizes, the true (unknown) magnitude of the difference or trend, and the true (unknown) variation. The rate at which we get it right, $(1 - \beta)$, is called the *power* of the test. Clearly, there is a trade-off between the two error rates: if we decrease α,

		The null hypothesis is (unknown to us)	
		TRUE	*FALSE*
Statistical decision	*Do not reject H_0*	✓	Type II error, at a rate β, in part a function of the experimental design
	Reject H_0	Type I error, at a rate α set at 0.05	✓

Figure (i) The four possible outcomes of a significance test in terms of accepting or rejecting the null hypothesis (*see text*).

then we increase β (all else being equal) and the power decreases.

Suppose we have two groups we are trying to discriminate with a *t*-test, each with the same normal distribution of measurements but with different mean values. If the means are far apart with scarcely overlapping distributions, then they are easy to distinguish by eye, but, more importantly, are likely to be different statistically (Fig. (ii)a). Move the mean values closer so that the distributions overlap substantially, and they become much harder to tell apart by eye, and are less likely to differ statistically (Fig. (ii)b). Reduce the variance so that once more the distributions hardly overlap, and once again they are easy to tell apart (Fig. (ii)c). These situations emphasise that the power of a test depends in part on the true (unknown) situation of the data, in this case the true difference between the means, and the true variance. How easy it is to tell the two groups apart is measured by a parameter called the *effect size*, which represents the 'distance' between H_0 (no difference) and H_1 (a significant difference).

The accuracy of our estimates of means and variances depend on the sample size we have chosen to take, and hence is under our control. Larger sample sizes provide more accurate estimates, and hence a given difference between two

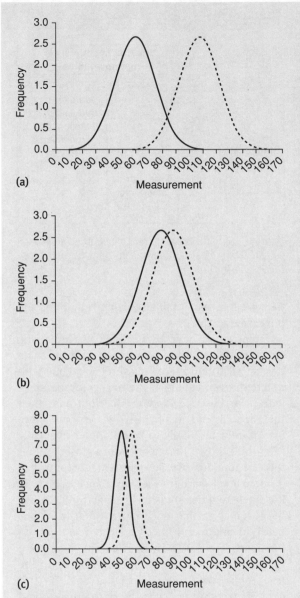

(a)

(b)

(c)

Figure (ii) Three different degrees of spread and separation in the distribution of two variables, leading to different degrees of discriminability (*see text*).

mean values is detected with greater power. Thus for a given statistical test, if we know a set of parameters (the Type I error rate α, the effect size, and the sample size), then we can calculate the power.

How much power is enough? What is an acceptable error rate for β? Like the significance level, α, this is essentially an arbitrary decision. The usual levels quoted in the literature are 80 per cent (i.e. $\beta = 0.20$), or (for those who believe that both Type I and Type II error rates should be the same) 95 per cent.

So what is a power test *for*? There are two main uses:

A priori power analysis

This is to ensure, *in advance*, that we are going to test our hypothesis adequately, that the experimental design we have chosen is actually capable of doing the job. We specify the power we want to have, and two of the three other parameters of α, effect size and sample size: given those, we can then calculate the third parameter. For a given design, the most obvious aspect under our control is the sample size, and hence a priori power tests are usually used to calculate how large the sample sizes should be in order to achieve a particular power. We specify α to be the conventional 0.05, and choose an effect size that indicates just how far away from H_0 we consider important enough to warrant attention. We can choose to pay attention to 'small', 'medium' or 'large' effects, depending on what theory, data or cost–benefit considerations tell us. The recommended values vary with the kind of test: for a t-test they are 0.2 (a small effect size), 0.5 (medium) and 0.80 (large); for ANOVA they are 0.1, 0.25 and 0.4, respectively; and for correlation and regression, 0.02, 0.15 and 0.35, respectively (what is being calculated varies among these tests, so they are not directly equivalent). Given a particular value of effect size, we can then calculate the required sample size using an appropriate software program (such as the free-to-download G*Power – R can do some power tests but G*Power is easier and more comprehensive). G*Power can be downloaded from **www.psycho.uni-duesseldorf.de/aap/projects/gpower/**. When we do this (Fig. (iii)), the results are often a bit of a blow – for example, using a one-tailed t-test to detect a medium-sized effect with $\alpha = 0.05$ needs a sample size of 102!

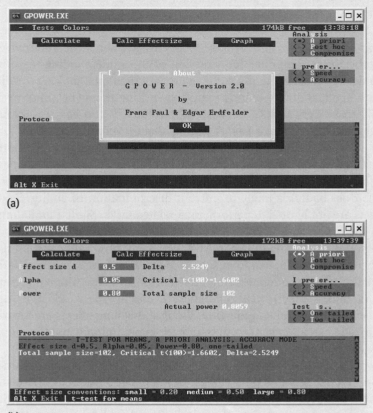

(a)

(b)

Figure (iii) Power calculation using G*Power. (a) Opening screen, (b) example of the sample size required for a one-tailed *t*-test to detect a medium-sized effect with $\alpha = 0.05$.

Post hoc power testing

Here, researchers retrospectively calculate the power of the test they have done. This is the kind of power test implemented in many statistical packages. An unfortunate tradition has built up that suggests that this kind of power test should routinely be used after a non-significant test and can be used to interpret the non-significant result – *it cannot* (*see* Hoenig & Heisey, 2001). Indeed, *post hoc* power tests are virtually useless since they tell us nothing more than the *p*-value itself gives.

we wanted to know whether the burden of a particular parasitic nematode increased with the body size of the host (e.g. a mouse). We might be tempted simply to assay worm burden and measure body size for a number of arbitrarily chosen host individuals, and then perform a correlation analysis. If we got a significant positive correlation coefficient, we might conclude that burden increased with body size. An unwelcome possibility, however, is that host body size correlates with age so that bigger hosts also tend to be older. If there is some age-related change in immune competence (e.g. older hosts are less able to resist infection), a positive correlation with size could arise that actually has nothing to do with host body size. In this case size is confounded with age, and simple correlation analysis cannot disengage the two. The best solution here would be to select different-sized hosts from a given age group so that the confounding effect is controlled for from the outset. Order effects are common confounding factors in many undergraduate projects. Testing animals in, say, treatment 1 first, then in treatment 2, then in 3, etc., confounds treatment with time. Animals may simply be tired by the time they get to treatment 3, so any difference in their response to the treatment could be due to that.

Floor and ceiling effects. Floor and ceiling effects arise when observational or experimental procedures are either too exacting or too undemanding in some way to allow a desired discrimination to be made. For example, looking for differences in mathematical ability among people by asking them the solution to $5 + 3$ is unlikely to be very fruitful because the problem is too easy; everyone will get the right answer straight away. A ceiling effect (everyone performs to a high standard) will thus prevent any differences there may be in ability becoming apparent. Conversely, if the same people were asked to solve a problem in catastrophe theory the odds are that no-one would be able to do it. In this case, a floor effect (everyone does badly) is likely to prevent discrimination of ability. Floor and ceiling effects are not limited to performance-related tasks. Similar limitations could arise in, for instance, histological staining. If a particular tissue requires just the right amount of staining to become discriminable from other tissues, the application of too little stain would result in everything appearing similarly pale (a floor effect) while too much stain would result in everything appearing similarly dark (a ceiling effect). Real differences in tissue type would thus not show up at the extremes of stain application. Floor and ceiling effects are clearly a hazard to be avoided and are well worth testing for in preliminary investigations.

Non-independence. Another common source of error in data collection and analysis arises from non-independence of data values. In many circumstances, there is a temptation to treat repeated measures taken from the same subject material as independent values during statistical analysis. As Martin & Bateson (1993) point out, this error arises from the misconception that the aim of a scientific observation or experiment is to obtain large numbers of measurements rather than measurements from a large number of subjects. The point is, of course, that obtaining additional measures from the same subject is not the same as increasing the number of subjects in the sample. An example of such an error would be as follows. Suppose an investigator wished to assess the average rate of nutrient flow in the phloem of a particular plant species. Setting up a preparation might be involved and time-consuming. To save effort, the investigator decides to take as many measurements as possible from each preparation before discarding it. As a result, there are 15 measurements from one preparation, 10 from another and 12, 16 and 5 from three more. To calculate the average, the investigator totals the measurements and divides by $n = 58$ (i.e. $15 + 10 + 12 + 16 + 5$). Of course, the measurements from each preparation are not independent; there may be something about the plant in each case that gives it an unusually high or low rate of nutrient flow relative to most of the plants in the population. Incorporating each measurement taken from it as an independent example of flow rate in the population as a whole is clearly going to bias the average upwards or downwards. The true n-size in the above example is five (the number of preparations), not 58. Measurements from each preparation should thus be averaged, or collapsed in some other way, to provide a single value for use in analysis. The fallacy of this kind of approach becomes obvious if we consider estimates of average plant height rather than nutrient flow rate. Few people would seriously measure the height of the same plant 16 times and regard these as independent samples of the height of the species concerned. The principle, however, is exactly the same in the flow rate example and is referred to as *pseudoreplication*.

The problem with non-independence is that it can operate at several different levels and sneak insidiously into analyses unless careful attempts are made to exclude it (*see*, for example, Boxes 3.4(a–b) and 3.5e). We have discussed it only at the level of the individual subject. However, depending on what is being measured, it could arise if, for example, related individuals are used as independent subjects or if plants grown on the same seed tray or animals kept in the same cage are used. We should thus be on our guard against it at all times. The golden rule is: one replicate equals one data point.

3.5   Testing predictions

Having highlighted some potential pitfalls, we must now bear them in mind as we return to our main observational examples and design experiments to test some of their predictions. We shall take one prediction from each.

Example 1

Plants and herbivores

E.g. Prediction 1A(ii) *Leaf damage by slugs will decrease the further up a plant that samples are taken.*

This predicts a negative trend between leaf damage and height up the plant. The assumption is that the height of a leaf off the ground influences its vulnerability to slugs. Two approaches immediately suggest themselves. We could conduct a survey in the field, measuring the height of leaves above the ground and scoring the amount of slug damage on each, or we could carry out an experiment, in the field or the laboratory, exposing leaves at different heights to slugs in a controlled environment. Either way, there is a formidable number of factors to take into account if we're to get a sensible outcome.

The most obvious is that our exploratory samples came from a wide range of plant species. At the very least there are likely to be confounding effects of species-specific attributes such as the presence of distasteful toxins or other deterrents. A first consideration in a field survey, therefore, might be to select plants of a similar range of heights within each of several species. This would ensure that the confounding effects of height and species were removed, but on its own it would still not be enough for a robust comparison. A major uncontrolled factor remains: the prevalence of slugs. Different species of plant are likely to occupy different habitats, some of which are more suitable for slugs than others. Another potential confounding effect may therefore need to be removed. To check, we could carry out a simple census of slug populations

around our subject plants. If numbers did not differ significantly, we could happily ignore them. If they did differ, however, we should need some way of taking them into account. One way would be to weight the recorded damage by the observed prevalence of slugs before analysis. We could analyse the relationship between weighted damage and height as a trend, using a correlation or regression analysis, but an alternative approach, which would allow us to look in more detail at the effect of plant species, would be to use two-way analysis of variance. A two-way analysis of variance, with height of leaves above the ground (low, medium, high) and plant species as the two levels of grouping, would reveal the separate main effects of height and plant species, but also the (very likely) interaction between them. The interaction term would probably be very important here because structural and developmental differences between species will almost certainly influence the effects of position up the plant. An alternative way to take slug prevalence into account in a parametric analysis of variance would be to include it as a covariate, a constant interval measurement whose effect on the data can be controlled for within the analysis of main effects and interaction.

If height did emerge as a significant predictor of slug damage, it would, of course, lead to further questions. Is the height effect due to slugs being unwilling to climb beyond a certain height? Is it due to leaves further up being tougher or more noxious? Is it due to taller plants having greater gaps between successive leaves so discouraging further ascent by slugs? Any height × species interaction might offer a clue to some of these (and other) possibilities by highlighting species characteristics that increase or decrease the height effect. An easier way to get at them, however, might be to do some laboratory experiments.

A laboratory study would attempt to control things more tightly at the outset. One approach might be to cultivate individual plants of some of the species sampled so that they were of similar height, and the important morphological characteristics, such as the number and spacing of leaves, were, as far as possible with different species, standardised. Plants of different species could then be arranged randomly or in a regular, alternating pattern on a bench, so that any systematic confounding of position and species was avoided, and each plant was exposed to the same number of similarly sized slugs for a set period, say overnight, and then scored for leaf damage. We might be tempted simply to catch a few slugs in the field and use those. However, this would be unwise. Freshly caught slugs would enter the experiment with an unknown feeding history. We would know neither their level of hunger, nor what they had recently been feeding on. Both factors could introduce unwelcome bias into the experiment or even cause a floor effect. The best thing to do would be to bring slugs into the laboratory well before the experiment (or culture them in the laboratory), feed them all on the same material (a combination of the plant species to be tested) and deprive them of food for a short time (e.g. 12 hours) prior to testing. All slugs would then be standardised for feeding experience and hunger.

The design above would allow us to assess the effect of leaf height on damage and, if we chose to observe slug activity (directly or using, for example, time-lapse photography), we might be able to conclude something about how any effect came about. Suppose we found that higher leaves did indeed sustain less damage, but that this was due not to slugs failing to get up to them but to slugs feeding for a shorter time when they did get there. Two possible explanations might be: (a) slugs were nearly satiated by the time they reached the higher leaves or (b) higher leaves are less palatable. An easy way to test for the latter would be to present slugs with standard-sized discs of material cut from leaves at different heights. We could choose three heights (high, medium and low) on different, but standard-sized, plants. If higher leaves are less palatable, discs cut from them would sustain less damage within the test time.

<table>
<tr><td>

Example 2

Hosts and parasites

</td><td>

E.g. Prediction 2A *Parasite burdens will increase with host testosterone levels.*

This prediction derives from the hypothesis that reproductive hormones might influence susceptibility to infection. Both sex and stress hormones are known to affect the immune system, often in concert, though their effects on resistance to parasites are very variable. Prediction 2A is based on the observation that adult voles in the samples generally had greater parasite burdens than juveniles, and males greater burdens than females (Fig. 2.1). Since the difference between age classes is much more pronounced in males, testosterone becomes a plausible candidate for driving the effects of age and sex on parasite burdens. As with the seemingly simple prediction about leaf height and herbivore damage, however, much needs to be thought about in testing whether there is a connection.

</td></tr>
</table>

We could start by taking some animals from the field and assaying their parasite burdens and testosterone levels. This could be done non-destructively by taking blood samples and faeces for blood and gut parasites, and inspecting the fur for ticks, fleas and other ectoparasites. Circulating testosterone concentrations could be assayed from either the blood or the faecal samples. Since testosterone secretion is highly pulsatile, and we are interested in chronic effects, faecal samples might provide the more appropriate measure since they accumulate testosterone metabolites over a period. Depending on the degree of discontinuity of testosterone concentration across age and sex classes, and the extent to which it can be normalised, we could test for testosterone as a predictor of age and sex differences in parasite burden by including it as a covariate in a two-way analysis of variance. The variable of interest would be our measure of parasite burden and the levels of grouping sex and age class. If we first ran the analysis without testosterone and found significant age and sex effects (as Fig. 2.1 suggests we might), but then found that the effects disappeared and were replaced by a significant covariate effect of testosterone when the latter was included, we should have some evidence that testosterone was important in

generating our initial age and sex differences in parasite burden. If the distribution of testosterone values did not permit this, we could instead test for differences in testosterone between age and sex classes using non-parametric analysis of variance and seek correlations between testosterone levels and parasite burden within classes.

Such analyses might tell us something, but it would be limited. One of the main problems is that simply taking animals from the field and testing for associations between hormone levels and parasites does not allow us to say anything about cause and effect. Testosterone may well influence parasite burden, but equally both may correlate with something else, such as level of social activity, which affects exposure to infection. The association would then be an artefact of testosterone levels and parasite burden being linked to exposure. A further problem is that testosterone may covary with circulating levels of corticosterone, a glucocorticoid 'stress' hormone that also influences the immune system. The best way to control for these potentially confounding effects is either to manipulate levels of testosterone and corticosterone directly, by injection or slow-release implants, and then see what happens when animals are given a controlled infection with a known parasite, or to monitor spontaneous levels of the hormones over a period and challenge in the same way. In order to use either approach, however, animals would need to be cleaned of existing parasites and given a period of acclimation before any experiment. Corticosterone plays a role in the immune response to many infections, particularly gut helminths (worms), and any residual infection is likely to compromise investigations of hormonal effects on resistance. The best approach to start with would probably be to monitor spontaneous levels of testosterone followed by challenge. While manipulating levels experimentally allows selective control of individual hormones, it can also cause unwanted side effects and disrupt the delicate interactions between physiological systems that underpin hormonal effects on resistance. It may also be necessary to remove relevant endocrine tissue surgically or by chemical ablation to prevent spontaneous secretion affecting the control of circulating levels. Such manipulations might thus be better as a follow-up to test conclusions arising from the more observational approach. The assumption in the latter, of course, is that differences in spontaneous circulating levels of hormone will predispose individuals to correspondingly different degrees of resistance. We should then look for an association between hormone levels prior to infection and the subsequent severity of the infection. Regression analysis would be the obvious candidate, perhaps using a suitable multivariate model to take the effects of both testosterone and corticosterone levels into account simultaneously.

Example 3

Nematodes and pollutants

E.g. Prediction 3B *Species present at unpolluted sites but missing from polluted sites will show greater mortality when exposed to pollutants.*

This prediction assumes that pollution is causally responsible for the absence of certain species from polluted samples. While this seems simple enough to test, we might want to explore its basis a little more before embarking on a set of experiments.

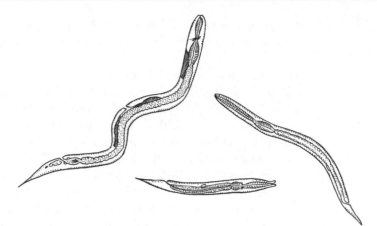

The most obvious point is that, with samples from only three sites, pollution status and site are confounded. Particular species may be present or absent at a given site purely by chance, or because sites happened to differ in some other important respect (e.g. interstitial water content of the soil) that affected their viability for different species. Ideally, therefore, we should first replicate our samples within site categories by choosing a number of sites, say four to six, of each kind (unpolluted, polluted with heavy metals, polluted with organophosphate) across which there is some variation in other environmental features. Species that are consistently absent from polluted sites in these samples would provide a better basis for further investigation.

One way forward might then be to collect or culture representative nematode species that are present only at unpolluted sites and expose them, say in standardised Petri dish cultures, to representative concentrations of heavy metal or organophosphate pollutant, not forgetting a suitable control (e.g. distilled water). Each treatment might be replicated half a dozen times. One-way analysis of variance of mortality by treatment would then reveal any significant effect due to the experimental pollutants. Significantly greater mortality in the two polluted treatments would be evidence in favour of a direct impact of pollution on species survival and thus presence/absence at particular sites. However, a lack of any effect would not necessarily rule out pollution as being responsible for the absence of certain species from polluted sites. Simply bathing adult worms in solutions and seeing if they die is a crude approach to say the least. Pollutants may work at any of a number of points in the worms' life cycle, perhaps reducing fecundity (the number of eggs produced) or the survival of larvae. Similar experiments could be performed to test these possibilities. More subtly, the effects may depend on particular environmental conditions, for example interactions between pollutants and other chemicals in the soil. This may be the case even though the simple experiment above showed a mortality effect; bathing worms in raw pollutant may kill them, but this may not be the way they are killed by pollution in the field. More complex experimental treatments, simulating patterns of exposure in the soil, might thus be called for.

| Example 4 |
| Crickets |

E.g. Prediction 4C *Encounters will progress further when opponents are more similar in size and it is more difficult to judge which will win.*

This prediction derived from observing that encounters between male crickets followed an apparently escalating pattern from chirping and antenna-tapping to out-and-out fighting and that, on the whole, bigger crickets tended to win. A possibility, therefore, is that progressive escalation reflects information-gathering about the relative size of an opponent and the likelihood of winning if the encounter is continued. If relative size is difficult to judge, as when two opponents are closely matched, the likelihood of winning cannot be judged in advance and the only way to decide the outcome is to fight. The prediction is thus of a negative trend between degree of escalation and the relative size of opponents: degree of escalation should increase with decreasing difference in size.

At first sight, this seems easy enough to test using a Spearman rank correlation or regression analysis. However, we first need some way of measuring degree of escalation. So far all we have are behavioural descriptions – chirping, antennating, fighting, etc. – from which we have inferred levels of escalation. Somehow we must put numbers to these. It is clear that we cannot put the behaviours on some common constant interval scale; we cannot, for instance, say that antennating is twice as escalated as chirping and fighting ten times as escalated. The easiest thing is simply to rank them. Thus what we assume to be the lowest level of escalation, say chirping, takes a rank of 1 and the highest level a rank of n, where n is the number of levels we decide to identify. We can then use the ranks of 1 to n as the y-values in our trend analysis.

To obtain our x-values, we must decide on a suitable measure of size. Ideally, the measure should be reliable and repeatable within and between individuals; measuring the size of the flexible abdomen, for instance, might not be a good idea because this could vary with food and water intake and thus vary from one encounter to another. It would be better to measure some component of the hard exoskeleton, e.g. the length of the long hind leg or the width of the thorax, which will not vary over the time course of observations. Of course, in our analysis, we are interested in a measure of the *relative* size of opponents, so our x-values must be some measure of relative size. The most obvious might be the *difference* in size between opponents. However, it is not hard to see why this would be inadequate. Suppose we observed two crickets of thorax widths 7.5 and 8.5 mm respectively. Suppose we observed another pair of thorax widths 6.5

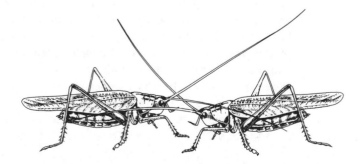

and 5.5 mm. In both cases, opponents differ by 1 mm and would score the same on a simple difference measure. In the first case, however, 1 mm is only 6 per cent of the combined width measures; in the second it is 8 per cent. A 1 mm difference may thus create a greater asymmetry in the likelihood of winning in the second case than in the first. As a result it would be better to use a ratio rather than a difference scale on the *x*-axis, e.g. size of bigger opponent/size of smaller opponent. However, ratios have strange statistical properties, and so we might want to use the ratio in logarithmic form, i.e. log(larger) – log(smaller).

Having decided on our measures, we can now plan observations. Since we are looking for a trend, we want to end up with pairs of *x*- and *y*-values. One way we might proceed is to put a number of individually marked males into a sand-filled arena and record all encounters over, say, 20 minutes, noting the males involved and the highest level (on our scale of 1 to *n*) to which each encounter progressed. One problem with this approach, however, is that some males would interact more than once. The pairs of *x*- and *y*-values arising from each repeat encounter could not be used independently because body size ratio would be confounded with pair of opponents and escalation levels might be influenced by the males' past experience of each other. It would therefore be better to arrange encounters between different pairs of males to provide independent replicates of a range of size ratios. Since we are using ordinal (rank) measures of *y* and selected ratios as *x*, and we have no reason to expect a linear relationship, we should test for significance in our trend using a Spearman rank correlation rather than regression.

3.6 Refining hypotheses and predictions

The discussions above do two things. First they give some simple indications as to how to set about testing particular predictions. Second, they show that the outcomes of such tests differ in the extent to which they increase or decrease our confidence in the hypothesis from which the prediction derives. Thus, for example, a failure to find an association between leaf damage and height up a plant in a field survey (Prediction 1A(ii)) would not greatly undermine our confidence in the hypothesis (1A) that leaf damage reflected availability to slugs. A host of factors besides height up the plant is likely to affect attack by slugs. Prediction 1A(ii) is thus a very restrictive test of Hypothesis 1A. The important point, however, is that the process of testing Hypothesis 1A does not stop there. It isn't abandoned just because one rather simplistic prediction did not work out. Instead the predictions are refined, gradually ruling out confounding factors. We saw this in the plants and herbivores example, as the suggested investigations used laboratory experiments to test for effects of height by controlling for changes in the size, spacing and palatability of leaves further up the plant.

Refinement also takes place in the opposite direction. A prediction that *is* borne out does not necessarily offer direct support for its parent hypothesis. The test of Prediction 3B (nematodes and pollutants) is a good example. This predicts that the species of nematode absent from polluted sites will die when exposed to pollutants in the laboratory. It derives from the general hypothesis

(3A) that pollution reduces species diversity. The finding that pollutants kill these species when they are bathed in them in Petri dishes does not necessarily mean that direct susceptibility to pollutants is the reason for reduced species diversity at polluted sites, or even the absence of those particular species from such sites. Their susceptibility in the laboratory may reflect a more general susceptibility to stressors. A variety of chemicals, quite unrelated to the environmental pollutants in question, might have a similar effect and this should certainly be tested. The absence of a particular species from polluted sites may reflect an interaction between effects of pollutants and competitive ability, with absence ultimately being due to competitive exclusion by other species rather than mortality through pollution. A more complex experimental design comparing effects on putatively robust (present at polluted sites) as well as putatively susceptible species would increase confidence in a mortality effect if differential mortality of susceptible, but not robust, species emerged.

Although we have presented hypotheses and their predictions in a rather cut-and-dried fashion through this book, it is clear that there is really considerable fluidity in both. The relationship between hypothesis, prediction and test is a dynamic one and it is through the modifying effects of each on the others that science proceeds.

Box 3.15　The principles of statistical analysis with R

As soon as you start to get to grip with the statistical analysis of data, you will see that the way R does it embodies a general approach that is important to know about. Once grasped, most analyses become simpler and rather straightforward, not harder and more complex.

The main objective is to find the values of the parameters of a model that produce the best fit to the data. Apart from coding errors and possible outliers, the data are 'sacrosanct, [telling] us what actually happened in a given set of circumstances' (Crawley, 2007: 324). Our job is to find the best model to fit to the data, i.e. the minimal sufficient model (see below) that results in the lowest amount of unexplained (residual) variation.

What are the main principles?

Explore the data via plots
It is always a good idea to plot the data to see the patterns. R has some superbly easy ways of doing this, that take all the effort away. The routines are well worth getting to know.

> `plot(x)`	an 'index plot' of the data values against position in the sequence, useful for error checking.
> `hist(x, breaks=seq(-0.5,9.5,1))`	a frequency distribution of the values of x, whether integers or real numbers; you control the bins of the distribution through the `breaks` argument.
> `plot(x,y)`	gives scatterplots (x continuous) or boxplots (x is a factor)
> `points(x,y)`	adds more points
> `abline(lm(y~x))`	adds a straight line
> `lines(x,y)`	adds a curve line joining points
> `lines(lowess(x,y))`	adds a smoothed non-parametric trend

```
> sequence<-order(x)
> lines(x[sequence],y[sequence])   makes sure points are in order
```

The following routines draw panels of plots, every variable in a dataframe against every other variable (`pairs`), or a plot of y vs x for different ranges of z (`coplot`).

```
> pairs(dtfr)
> pairs(dtfr, panel=panel.smooth)
> coplot(y ~ x|z)
```

`Lattice` is a library for panels of plots of various types, a very powerful way of visualising the patterns in the data (see Zuur *et al.*, 2009: chapter 8):

```
> library(lattice)
> xyplot(y~x|f)
```

and similar multiple plots can be done with '`barchart`' (barplots) and '`bwplot`' (box-and-whisker plots). > `histogram(~x|A)` will plot multiple frequency distributions of x split by levels of A.

Most or all of these routines can take a set of graphical arguments. The full set can be seen by typing `?par` in R: `xlab=c("x-axis label")`, `ylab`; `lty=1` (solid line) or `2` (dashed line); `pch=0` (open square), `1` (open circle), `3` (open triangle), `4` (cross), `20` up (filled versions); `col=1` (black), `2` (red), `3` (blue), etc; `cex` = relative size of symbols. The common ones are listed in Appendix IV.

General Linear Models

After a bit of experience in using the R commands for differences (`aov`) and trends (`lm`), you will rapidly see that these are simple subsets of General Linear Models (the command `glm`), where there are simply predictors of two types: factors (differences, where the estimated parameters are the mean values of each level of the factor), and continuous variables (trends, where the estimated parameter is a slope). Factors come in two types: fixed (where levels represent only themselves) and random (where levels are intended to represent the universe of possible levels, and hence are generalisable to that universe).

The output from a glm takes a bit of getting used to, since it looks quite different. Take, for example, the one-way ANOVA data of Box 3.5a. Using aov:

```
> m1<-aov(bbee~treat)
> summary(m1)
          Df  Sum Sq Mean Sq F value  Pr(>F)
treat       3   80.28 26.7586  3.3131 0.02163 *
Residuals 156 1259.95  8.0766
```

whereas using glm:

```
> m2<-glm(bbee~treat)
> summary(m2)
glm(formula = bbee ~ treat)
Coefficients:
            Estimate Std. Error t value Pr(>|t|)
(Intercept)  12.4785     0.4493  27.770  < 2e-16 ***
treatleaves  -1.3425     0.6355  -2.113  0.03623 *
treatnormal  -1.2417     0.6355  -1.954  0.05248 .
treatstones  -1.9500     0.6355  -3.069  0.00254 **
--
```

```
(Dispersion parameter for gaussian family taken to be 8.076575)
Null deviance: 1340.2 on 159 degrees of freedom
Residual deviance: 1259.9 on 156 degrees of freedom
AIC: 794.24
```

In the glm output, you get one line per parameter estimated; apart from the 'intercept', you get as many parameters as there are degrees of freedom in the design. Here there are three degrees of freedom, and hence three parameters. Notice that the 'dispersion parameter' is the residual variance (the mean square). Unlike in regression, where the intercept is the overall mean, here the 'intercept' is the mean value for the first group (here, the control). The other parameters in a one-way design are differences of each group from the 'intercept' (the mean for the first group): their s.e.s are of the differences, not the means (which is why they are bigger).

In two-way designs it gets a bit more complex. Let's redo the two-way design of Box 3.9a as a glm:

```
> m1<-aov(enzyme~parasitism*diet)
> summary(m1)
                Df Sum Sq Mean Sq F value    Pr(>F)
parasitism       1 1.6504 1.65039 26.4377 8.053e-07 ***
diet             1 0.0166 0.01661  0.2660   0.60675
parasitism:diet  1 0.3358 0.33581  5.3793   0.02167 *
Residuals      156 9.7384 0.06243

> m2<-glm(enzyme~parasitism*diet)
> summary(m2)
glm(formula = enzyme ~ parasitism * diet)
Coefficients:
                                   Estimate Std. Error t value Pr(>|t|)
(Intercept)                         9.86550    0.03950 249.728  < 2e-16 ***
parasitism-unparasitized            0.29475    0.05587   5.276 4.35e-07 ***
diet-peas                           0.11200    0.05587   2.005   0.0467 *
parasitismunparasitized:dietpeas   -0.18325    0.07901  -2.319   0.0217 *
---
(Dispersion parameter for gaussian family taken to be 0.06242563)
Null deviance: 11.7412 on 159 degrees of freedom
Residual deviance: 9.7384 on 156 degrees of freedom
AIC: 16.205
```

In the ANOVA there are three d.f., one each for the main effects and one for the interaction. In the glm there are also three d.f., but they seem completely different. Again there is an 'intercept' and then three parameters, and again they are differences. The intercept is again the first group, here the parasitised-beans group. The first parameter is for column 2 (unparasitised), and is the difference between the 'intercept' and both groups in column 2 (unparasitised groups). The second parameter is the difference between the 'intercept' and both groups in row 2 (the peas diet). The third is the difference between the observed mean of the unparasitised peas group, and the prediction based on intercept + col2 + row2; this is the interaction, the non-additivity of the column and row effects. Notice again how the 'dispersion parameter' is the residual variance (the mean square).

Generalised Linear Models

General Linear Models allow great flexibility in the predictors, but all response variables are supposed to have normally distributed residuals. The huge advance in recent years has been to extend this range to a set of different residual distributions in Generalised Linear Models: Poisson, binomial, gamma, etc. (see Box 3.1a). This has meant that many analyses that previously had to be done using non-parametric methods can now be done using the much more flexible, powerful and complex parametric models. The impact of this on statistical analysis can hardly be exaggerated. In R the various distributions are very easily invoked:

```
> glm(counts ~ parasitism * diet, family = Poisson)
```

Crawley (2007) gives a full account of these kinds of models.

Model checking

Since the idea is to fit the best and simplest model to the data, we need to check whether we have the correct model. R has some very simple diagnostics to help. Usually once the model has been fitted and saved (into 'model', for example), we just type:

```
> plot(model)
```

Table (i) gives the assumptions to be tested for each error distribution, and how to check them.

Parsimony and order effects

The principle of parsimony is the idea that science seeks the simplest explanation consistent with the facts. In statistics, this means obtaining the minimal set of predictors, getting rid of any that fail to predict significant amounts of variation in the data. Thus we start with the full set of predictors and gradually reduce them according to a selected criterion until we are left with the minimal sufficient model containing just significant predictors. One criterion commonly used is called the AIC, or Akaike Information Criterion.

The more parameters there are in a model, the better the fit (i.e. the more of the variation in the data are accounted for by the model, and the less the variation in the residual, or error, variation – the bit unaccounted for by the model). Of course, this means that we could predict the data perfectly if we had one parameter for every data point. Clearly this would not be a very good model. So we need a criterion where these two things are traded off against each other – the number of parameters (= bad) against the proportion of the variation accounted (= good). The AIC does just this: known as penalised log-likelihood, there is a penalty added for each parameter in the model. Thus we prefer models with lower AIC values, because these have the best tradeoff between the number of parameters and the proportion of the variation accounted for: they are a better fit. Thus we choose to delete predictors if this reduces the AIC of the subsequent model. This is the principle automated by the R command 'step'.

```
> step(model)
```

One of the things that dispirits students the most, causing them more grief almost than anything else, is the fact that the order of the predictors matters in many statistical models. Students are dismayed by the fact that if they change the order of the predictors, then all the significances change. This happens for three reasons:

- The design of the experiment is unbalanced (i.e. replication is not the same across all groups); many kinds of investigation have no control over replication, so this is not something that can be fixed in many cases.
- Predictors are correlated with one another (i.e. are non-orthogonal).
- Because of the nature of regression, which assumes no error in the x-values; since a second predictor (continuous covariate) is trying to account for the residual variation left over from fitting the

Table (i) How to check in R whether your model is correct

checking	y	x	for	expected pattern	what to do with unexpected patterns
use plot(model1) to get					
	#1 residuals	fitted values	heteroscedasticity, non-linearity	no pattern	if variance increases with fitted values, use sqrt(response), log(response), or Poisson errors
	#2 standardized residuals	normal scores/quantiles	normality (the q-q plot)	straight line (especially at ends)	use different errors
	#3 sqrt(standardized residuals)	fitted values	heteroscedasticity	no pattern	
	#4 standardized residuals	leverage (a measure of each point's influence on the model)		no pattern	
then also plot:	response	each predictor	outliers, points with high leverage	no outliers	remove points or reduce influence by weighting
	residuals	each predictor	heteroscedasticity, non-linearity	no pattern	for one predictor, if variance increases or if curved, transform the predictor using sqrt or log
	residuals	time of collection	temporal autocorrelation	no pattern	

Error Distribution	link	When used	Parameters of distribution	estimated parameters	assumptions	If assumptions not valid
normal	identity	measures, ranks, large-number integers	mean, var	means, slopes	constant variance (use #1), normal errors (use #2)	transform response, or use different link (eg log), or different error structure
gamma (positive skew – ie with a long right-hand tail)	reciprocal	measures, ranks, large-number integers	shape α, rate β; mean = $\alpha\beta$, var = $\alpha\beta^2$. Collapses to exponential when $\alpha = 1$	means, slopes	constant coefficient of variation (variance increases faster than linearly with mean); normal standardised residuals (use #2)	
Poisson	log	counts	mean	means of log(response), slopes of log(response) vs predictor	variance = mean (residual deviance = residual df, ie scale parameter = 1); normal standardized residuals (use #2)	if variance > df (overdispersed), use quasipoisson, or use F-tests rather than Chi, or transform with normal errors, or use negative binomial errors
binomial	logit[1]	0/1 data	probability of success	means/slopes of logit-transformed response; back-transform to proportions via $p = 1/(1 + (1/e^z))$	variance < mean (residual deviance < residual df, ie scale parameter < 1); normal standardized residuals (use #2)	no such thing as overdispersion of a 0/1 response variable; choose the link that minimises the residual deviance. If non-linear, can use GAM smoothing functions
binomial	logit[1]	proportions (N out of N+M): use cbind to create a two-column matrix from vectors of N and M	probability of success	means/slopes of logit-transformed response; back-transform to proportions via $p = 1/(1 + (1/e^z))$	variance a $\cap$-shaped function of the mean (bounded between 0 and 1); normal standardized residuals (use #2)	if variance > df (overdispersed), use F-tests rather than Chi, or quasibinomial
negative binomial: use function glm.nb	log[1]	counts of aggregated organisms	mean, k (aggregation parameter)	means of log(response), slopes of log(response) vs predictor	variance >> mean (residual deviance >> residual df, ie scale parameter >> 1); normal standardized residuals (use #2)	
quasi-likelihood		when you are not sure which distribution to use				
special	specify using quasi					

1. Sometimes it can also be better to use a log-log link

first, it is obvious that fitting them the other way round generates a completely different set of residuals for the second predictor.

Deleting predictors from the maximal model is the safest way to proceed.

Effect size

Using R and glms encourages the investigator to concentrate on the magnitude of the effect of a predictor, rather than its significance. An effect can be highly significant yet biologically irrelevant, and conversely only just significant yet biologically highly relevant.

Crawley (2007) is an excellent guide to the use of R for General and Generalised Linear Models, and explains very clearly the principles enumerated here.

3.7 Summary

1. The form of the predictions derived from hypotheses dictate the design of the experiments and observations needed to test them. As a result of testing, hypotheses can be rejected, provisionally accepted or modified to generate further testable hypotheses.

2. Decisions about experimental/observational measurement and the confirmatory analysis of such measurements are interdependent. The intended analysis determines very largely what should be measured and how, and should thus be clear from the outset of an investigation.

3. Some kind of yardstick is needed in confirmatory analysis to allow us to decide whether there is a convincing difference or trend in our measurements (i.e. whether we can reject the null hypothesis of no difference or trend). The arbitrary, but generally accepted, yardstick is that of statistical significance. Significance tests allow us to determine the probability that a difference or

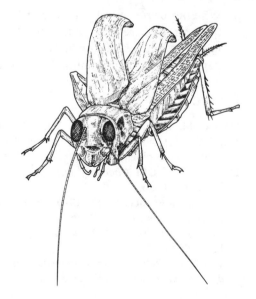

trend as extreme as the one we have obtained could have occurred purely by chance. If this probability is less than an arbitrarily chosen threshold, usually 5 per cent but sometimes 1 or 10 per cent, the difference or trend is regarded as significant and the null hypothesis is rejected.

4. Different significance tests may demand different attributes of the data. Parametric and non-parametric tests differ in the assumptions they make about the distribution of data values within samples and the kinds of measurement they can cope with. Tests can also be used in specific/one-tailed or general/two-tailed forms depending on prior expectations about the direction of differences or trends.

5. Basic tests for a difference include chi-squared, the *t*-test, the Mann–Whitney *U*-test and one- and two-way analysis of variance. Each has a number of requirements that must be taken into account, and care is needed not to make multiple use of two-group difference tests in comparing more than two groups of data.

6. Basic tests for a trend include Pearson product moment and Spearman rank correlations and regression analysis. Correlation is used when we merely wish to test for an association between two variables. Regression is used when we change the values of one variable experimentally and observe the effect on another to test for a cause-and-effect relationship between two variables. Regression analysis yields more quantitative information about trends than correlation but makes more stringent demands on the data.

7. Multifactor and multivariable forms of difference and trend analyses allow the effects of several different variables to be analysed together within tests.

8. Testing predictions requires careful thought about methods of measurement, experimental/observational procedure, replication and controlling for confounding factors and floor and ceiling effects.

9. Whether or not a test supports or fails to support a prediction, the implications for the parent hypothesis need careful consideration. A hypothesis does not necessarily stand or fall on the outcome of one prediction; everything depends on how discriminating the prediction really is.

References

Crawley M. J. (2007) *The R book*. Wiley, London.

Day, R. W. & Quinn, G. P. (1989) Comparisons of treatments after an analysis of variance in ecology. *Ecological Monographs* **59**, 433–463.

Dytham, C. (2010) *Choosing and using statistics: a biologist's guide*. 3rd edition. Blackwell, Oxford.

Gaines, S. D. & Rice, W. R. (1990) Analysis of biological data when there are ordered expectations. *American Naturalist* **135**, 310–317.

Grafen, A. & Hails, R. (2002) *Modern statistics for the life sciences*. Oxford University Press, Oxford.

Hawkins, D. (2005) *Biomeasurement: understanding, analysing, and communicating data in the biosciences*. Oxford University Press, Oxford.

Hoenig, J. M. & Heisey, D. M. (2001) The abuse of power: the pervasive fallacy of power calculation for data analysis. *American Statistician* **55**, 19–24.

Martin, P. & Bateson, P. (1993) *Measuring behaviour*, 2nd edition. Cambridge University Press, Cambridge.

Meddis, R. (1984) *Statistics using ranks: a unified approach*. Blackwell, Oxford.

Ridley, J., Kolm, N., Freckleton, R. P. & Gage, M. J. G. (2007) An unexpected influence of widely used significance thresholds on the distribution of *P*-values. *Journal of Evolutionary Ecology* **20**, 1082–1089.

Ruxton, G. D. & Colegrave, N. (2006) *Experimental design for the life sciences*, 2nd edition. Oxford University Press, Oxford.

Siegel, S. & Castellan, N. J. (1988) *Nonparametric statistics for the behavioral sciences*. McGraw-Hill, New York.

Sokal, R. R. & Rohlf, F. J. (1995) *Biometry*, 3rd edition. Freeman, San Francisco.

Zuur, A. F. Ieno, E. N. & Meesters, E. H. W. G. (2009) *A beginner's guide to R*. Springer, London.

4 Presenting information
How to communicate outcomes and conclusions

If your investigation has gone to plan (and possibly even if it hasn't), you will have generated a pile of data that somehow needs to be presented as a critical test of your hypothesis. Performing appropriate significance tests is only one step on the way. While significance tests will help you decide whether a difference or trend is interesting, this information still has to be put across so that other people can evaluate it for themselves. There are two reasons why we should take care how we present our results. The first is to ensure we get our message over; there is little point making a startling discovery if we can't communicate it to anyone. The purpose of our investigation was to test a hypothesis. We might conclude that the results support the hypothesis or that they undermine it. Whichever conclusion we reach we must sell it if we wish it to be taken seriously. Since scientists are by training sceptical, selling our conclusion may demand some skilful presentation and marshalling of arguments. The second reason is that we must give other people a fair chance to judge our conclusions. As we saw earlier, simply saying that some difference or trend is significant doesn't tell us how strong the effect is. It is important to present results in such a way that others can make up their own minds about how well they support our conclusions. In this chapter, we shall look at some conventions in presenting information that help satisfy both these requirements. We begin with some simple points about figures and tables.

4.1 Presenting figures and tables

We stressed earlier that it is usually not helpful to present raw data. Raw data are often too numerous and the information in them too difficult to assimilate for useful presentation. Instead, we summarise them in some way and present the summary form. We have already dealt with summary statistics in a general way when we discussed exploratory analysis; here we discuss their use in presenting the results of confirmatory analysis.

Although it is usually obvious that raw data require summarising, there can be a temptation to summarise everything, as if summary statistics or plots were of value in themselves. In confirmatory analyses, they are of value only to the extent that they help us evaluate tests of hypotheses. We thus need to be selective in distilling our results. Naturally, the summaries and forms of presentation that are most appropriate will depend on the type of confirmatory analysis. It is therefore easiest to deal with different cases in turn.

4.1.1 Presenting analyses of difference

Where we are dealing with analyses of difference, the important information to get across is a summary of the group values being compared. The form this takes will vary with the number of groups and levels of grouping involved. There are two basic ways of presenting a summary of differences: figures and tables. As we have argued previously, figures tend to be easier to assimilate than tables, even when the latter comprise summary statistics. However, tables may be more economical when large numbers of comparisons are required, or where comparisons are subsidiary to the main point being argued but helpful to have at hand. If tables are used, it is important that they present all the key summary information necessary to judge the claims they make. This usually means (a) summary statistics (e.g. means ± standard errors,[†] medians ± confidence limits) for each of the groups being compared, (b) the sample size (n) for each group, (c) test statistic values, (d) the probabilities (p-values) associated with the test statistic values and (e) an explanatory legend detailing what the table tells us. The test statistics and p-values can be presented either in the table itself or in the legend. The same information, of course, should be presented in figures except that the summary statistics are represented graphically (e.g. as bar charts) instead of as numbers, and information about sample sizes, test statistics and probability levels more conventionally goes in the legend (now usually called the *figure caption*) rather than in the figure itself. (Nevertheless, as long as it doesn't clutter the figure and detract from its impact, it can be very helpful to include statistical information within the figure, and we shall do this later where appropriate.)

Differences between two or more groups (with one level of grouping)
Here we are presenting the kinds of result that might emerge from a Mann–Whitney U-test, or a one-way analysis of variance. Suppose we have tested for a difference in growth rate (general prediction) between two groups of plants, one given a gibberellin (growth-promoting) hormone treatment, the other acting as

---

[†] While there are various forms of summary statistic, means ± standard errors are widely used because the mean of a set of values is an easy concept to grasp and because the standard error estimates the distribution of *means* within the population, which approaches a normal distribution as the sample size increases. It is thus usually legitimate to quote means ± standard errors as summary statistics even when the distribution of data *values* demands a non-parametric significance test.

an untreated control. The treated group contained 12 plants and the untreated group eight. Using a one-way analysis of variance for two groups, we discover a significant difference at the 0.1 per cent ($p < 0.001$) level between the groups, with treated plants growing to a mean (± standard error) height of 14.75 ± 0.88 cm during the experimental period, and controls growing to a mean height of 9.01 ± 0.63 cm. We could present these results as in Table 4.1a. Note the legend explaining exactly what is in the table.

Table 4.1a The mean height to which plants grew during the experimental period when treated with gibberellin or left untreated.

	Experimental groups		
	Treated	Untreated	Significance
Mean (± s.e.) height (cm)	14.75 ± 0.88	9.01 ± 0.63	$H = 12.22$,
n	12	8	$p < 0.001$

The significance column could be omitted from the table, in which case the test statistic and probability level should be given in the legend. The legend would now read:

Table 4.1a The mean height to which plants grew during the experimental period when treated with gibberellin or left untreated. H comparing the two groups $= 12.22$, $p < 0.001$.

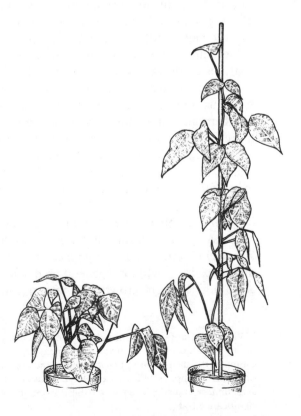

An alternative and frequently adopted convention in presenting significance levels is to use asterisks instead of the test statistic and probability numbers. In this case, different levels of probability are indicated by different numbers of asterisks. Usually * denotes $p < 0.05$, ** $p < 0.01$ and *** $p < 0.001$, but this can vary between investigations so it is important to declare your convention when you first use it. Using the asterisks convention, the table would now read as in Table 4.1b. Exactly the same forms of presentation, of course, could be used for comparisons of more than two groups. However, if we had used a U-test to test for a difference between two groups, we should now have to change to a one-way analysis of variance to avoid abuse of a two-group difference test (*see* Chapter 3).

Table 4.1b The mean height to which plants grew during the experimental period when treated with gibberellin or left untreated. ***, $H = 12.22$, $p < 0.001$.

	Experimental groups		
	Treated	Untreated	Significance
Mean ($\pm$ s.e.) height (cm)	14.75 ± 0.88	9.01 ± 0.63	***
n	12	8	

The presentation of means and standard errors (or medians and confidence limits) is appropriate whenever we are dealing with analyses that take account of the variability within data samples. In chi-squared analyses, however, where we are comparing simple counts, there is obviously no variability to represent. If we were presenting a chi-squared analysis of the number of plants surviving each of three different herbicide treatments and one control treatment, therefore, the table would be as shown in Table 4.2a. Table 4.2b shows an alternative presentation. If we wish to present our results as figures rather than tables, we can convey the same information using simple bar charts. Thus Table 4.1a can be recast as Fig. 4.1a. Similarly, Table 4.1b could be recast as Fig. 4.1b. For a comparison of three groups, say comparing the effectiveness of the lambda bacteriophage in killing three strains of *Escherichia coli* suspected of differing in susceptibility, the figure might be as shown in Fig. 4.2a.

In Figs 4.1a, b and 4.2a, we have assumed a one-way analysis of variance was used to test a general prediction (hence the test statistic H). If we had instead tested a specific prediction because we had an a priori reason for expecting a rank order of effect (e.g. gibberellin-treated plants would grow taller than untreated plants (Fig. 4.1), or strain B of *E. coli* would be most resistant and

Table 4.2a The number of plants surviving treatment with different herbicides.

	Experimental group				
	Herbicide 1	Herbicide 2	Herbicide 3	Control	Significance
Number of plants surviving	15	8	6	27	$\chi^2 = 19.3$, $p < 0.001$

Table 4.2b The number of plants surviving treatment with different herbicides. ***, $\chi^2 = 19.3$, $p < 0.001$.

	Experimental group				
	Herbicide 1	Herbicide 2	Herbicide 3	Control	Significance
Number of plants surviving	15	8	6	27	***

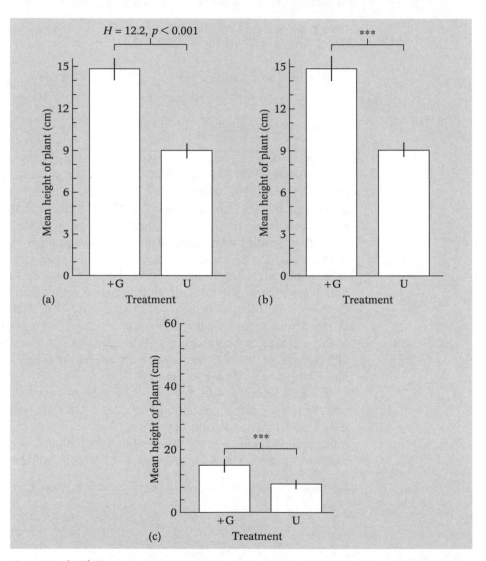

Figure 4.1 (a, b) The mean height to which plants grew during the experimental period when treated with gibberellin (+G, $n = 12$) or left untreated (U, $n = 8$). ***, $H = 12.2$, $p < 0.001$. Bars represent standard errors. (c) Figure 4.1b with a different scale. The mean height to which plants grew during the experimental period when treated with gibberellin (+G, $n = 12$) or left untreated (U, $n = 8$). ***, $H = 12.2$, $p < 0.001$. Bars represent standard errors.

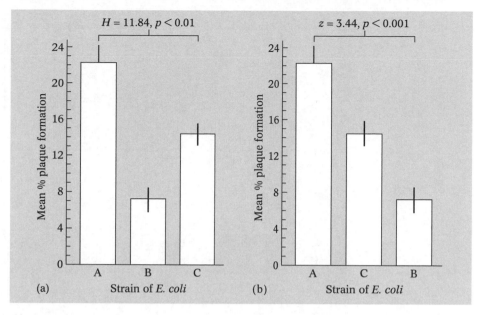

Figure 4.2 The mean percentage area of plaque (= bacterial death) formation by lambda bacteriophage on three strains (A–C) of *E. coli*. Bars represent standard errors. (a) Non-parametric analysis of variance testing a general prediction of difference between strains. $N = 8$ cultures in each case. (b) Non-parametric analysis of variance testing a specific prediction of difference between strains (A > C > B). $N = 8$ cultures in each case.

strain A least resistant to attack by the phage (Fig. 4.2)), we should recast the figures with groups in the predicted order and quote the test statistic z rather than H. Thus Fig. 4.2a could be recast as Fig. 4.2b.

For most people, bar charts like these convey the important differences between groups more clearly and immediately than equivalent tables of numbers. However, it is worth stressing some key points that help to maximise the effectiveness of a figure.

1. Make sure the scaling of numerical axes is appropriate for the difference you are trying to show. For instance, the impact of Fig. 4.1b is much reduced by choosing too large a scale (*see* Fig. 4.1c).

2. Always use the *same* scaling on figures that are to be compared with one another. Thus, Fig. 4.3a, b is misleading because the different scaling makes the magnitude of the bars look the same in (a) and (b). Using the same scale, as in Fig. 4.3c, shows that there is in fact a big difference between (a) and (b).

3. Make sure axes are numbered and labelled properly and that labels are easy to understand and indicate the units used. Avoid obscure abbreviations in axis labels: these can easily be ambiguous and misleading or unnecessarily difficult to interpret.

4. Axes do not have to start at zero. Presentation may be more economical if an axis is broken and starts at some other value. Thus, Fig. 4.4a could be recast as Fig. 4.4b with the break in the vertical axis indicated by a double slash.

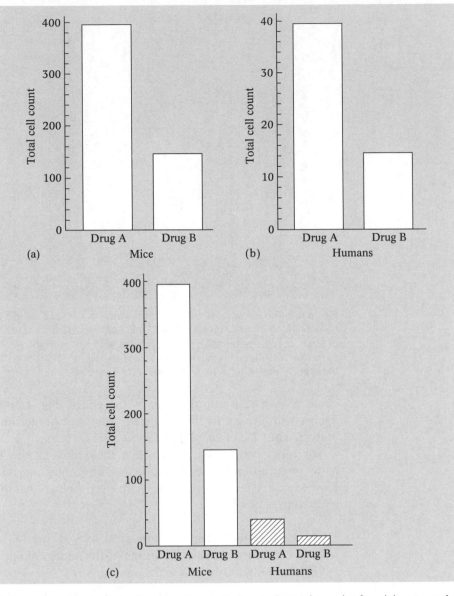

Figure 4.3 (a) The total number of T-helper cells in experimental samples from laboratory mice following administration of two different cytotoxic drugs (A and B). $N = 40$ samples for each drug treatment. (b) As Fig. 4.3a, but for experimental samples from humans. (c) The total number of T-helper cells in experimental samples from laboratory mice (open bars) and humans (shaded bars) following administration of two different cytotoxic drugs (A and B). $N = 40$ samples for each drug treatment and species.

5. Include indications of variability (standard errors, etc.) where appropriate. Also include sample sizes and p-values as long as these don't clutter the figure. If they do, put them in the legend.

6. Always provide a full, explanatory legend. The phrasing of the legend should be based on the prediction being tested, and the legend should include

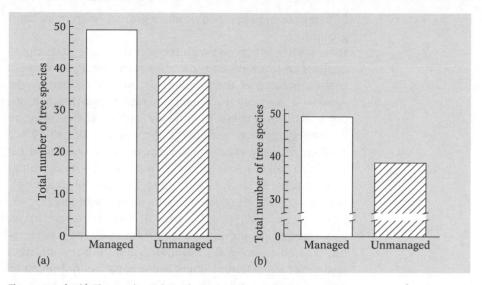

Figure 4.4 (a, b) The total number of species of trees bearing epiphytes in 2 km² study areas of rainforest in Bolivia where forests are managed economically (open bar) and unmanaged (shaded bar). $N = 1 \times 2$ km² area in each case.

any statistical information (sample sizes, test statistics, etc.) not included in the figure. Do not repeat information in both figure and legend, though. The legend should allow a reader to assess the information in the figure without having to plough through accompanying text to find more detailed discussion.

Differences between two or more groups (with more than one level of grouping)

Here we are concerned with the sort of results that might arise from a two-way analysis of variance or $n \times n$ chi-squared analysis. Presentation is now a little trickier because of the number of comparisons we need to take into account. A table is probably the simplest solution. For instance, suppose we had carried out a two-way analysis of variance looking at the difference in the frequency of accidental egg damage between three strains of battery hen maintained in three different housing conditions. Here we have two levels of grouping (strain and housing condition) with three groups at each level. The analysis tests for a difference between strains (controlling for housing condition), a difference between housing conditions (controlling for strain) and any interaction between the two levels of grouping (*see* Chapter 3). The best way to present the differences between groups within levels is to tabulate the summary statistics for each of the nine (3×3 groups) cells and include the test statistics in the legend. Thus testing for any difference between groups (i.e. not predicting a difference in any particular direction) might give the results shown in Table 4.3.

Table 4.3 The mean ($\pm$ s.e.) percentage number of eggs broken during the experimental period by three strains of battery hen (1–3) under three different housing conditions (A–C). Parametric two-way analysis of variance shows a significant effect of both strain ($F = 145.09$, d.f. = 2,27, $p < 0.001$) and housing ($F = 103.29$, d.f. = 2,27, $p < 0.001$) and a significant interaction between the two ($F = 58.76$, d.f. = 4,27, $p < 0.001$). $N = 4$ in each combination of strain and housing condition.

		Strain		
		1	2	3
	A	43.50 ± 2.32	1.25 ± 0.75	22.25 ± 2.14
Housing condition	B	38.75 ± 1.09	13.75 ± 1.38	16.50 ± 1.71
	C	6.25 ± 0.85	10.25 ± 1.80	6.25 ± 1.80

This analysis reveals significant effects of both strain and housing conditions on egg breakage. These are obvious from the summary statistics in the table: breakage in strains 1 and 3 is relatively high under housing conditions A and B but drops sharply in condition C. In contrast, breakage in strain 2 is highest in conditions B and C and lowest in A. Damage tends to be greater in types A and B housing than in type C. In addition to these main effects, however, there is also a significant interaction between strain and housing condition (*see* legend to Table 4.3) with the effect of housing differing between strains. Although interaction effects can also be gleaned from a table of summary statistics like Table 4.3, they can be presented more effectively as a figure; one of the levels of grouping constitutes the *x*-axis and the measure being analysed is the *y*-axis. The relationship between the measure and the *x*-axis grouping can then be plotted for each group in the second level. Figure 4.5 shows such a plot for the interaction in Table 4.3. The lines in the figure, of course, simply indicate the groups of data: they are in no way comparable with statistically fitted lines. Full details of the analysis are given in the legend because such a figure would not normally be presented as well as the summary table since it repeats information already given

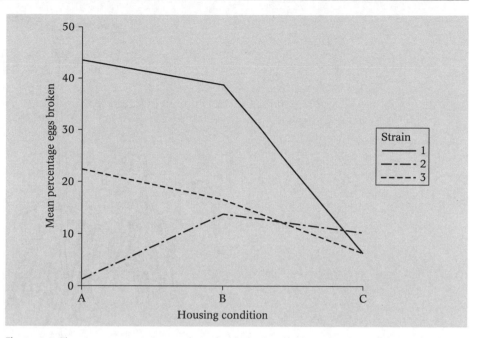

Figure 4.5 The mean percentage number of eggs broken by three strains of battery hen (solid, strain 1; dash/dot, strain 2; dotted, strain 3) in three different housing conditions (A–C). Parametric two-way analysis of variance showed a significant effect of both strain ($F = 145.09$, d.f. = 2,27, $p < 0.001$) and housing condition ($F = 103.29$, d.f. = 2,27, $p < 0.001$) and a significant interaction between the two ($F = 58.76$, d.f. = 4,27, $p < 0.001$). $N = 4$ in each combination of strain and housing condition.

in the table. The graph should have the s.e.s plotted with the mean values, but it does tend to become very cluttered. Judgement is required as to whether a table or a figure is the best presentation. From Fig. 4.5 it is clear that, while all three strains show differences in egg damage across housing conditions, the direction and degree of decline are different in different strains. This implies that the effect of housing condition varies with strain, which is what is meant by an interaction between housing condition and strain.

With an $n \times n$ chi-squared analysis, we are just dealing with total counts in each cell so there are no summary statistics to calculate and present. The simplest presentation is thus an $n \times n$ table with each cell containing the observed and expected values for the particular combination of groups (the expected value in each cell usually goes in brackets). Table 4.4 shows such a presentation for a

Table 4.4 The number of seeds germinating in a tray in relation to temperature (low, 5 °C; high, 25 °C) and soil type. Expected values in brackets. $\chi^2 = 14.38$, d.f. = 1, $p < 0.001$.

	Number of seeds germinating	
	In clay soil	In sandy soil
Low temperature	40 (57.97)	100 (82.03)
High temperature	131 (113.03)	142 (159.97)

chi-squared analysis of the effects of temperature and soil type on the number of seeds out of 150 germinating in a seed tray.

4.1.2 Presenting analyses of trends

Presenting trend analyses is rather simpler because, in most cases, a scattergram with or without a fitted line is the obvious format. When it comes to more complicated trend analyses that deal with lots of different measures at the same time, it is usually possible to present the various significant relationships as so-called *partial* correlation or regression plots. These can be selected within the regression analysis procedures of R, for example. Alternatively, summary tables rather than figures may be necessary.

Presenting a correlation analysis

Since correlation analysis does not fit a line to data points, presentation consists simply of a scattergram, though depending on how we have replicated observations this may include some summary statistics (*see below*). Information about test statistics, sample sizes and significance could be given in the figure, but it is more usual to include it in the legend. Thus Fig. 4.6a shows a plot of the number of food items obtained by male house sparrows (*Passer domesticus*) in relation to their dominance ranking with other males in captive flocks of six (rank 1 is the most dominant male, which tends to beat all the others in aggressive disputes, and rank 6 is the least dominant, which usually loses against everyone else). In this case, observations were repeated for three sets of males so there are three separate points (*y*-values) for each *x*-value in the figure.

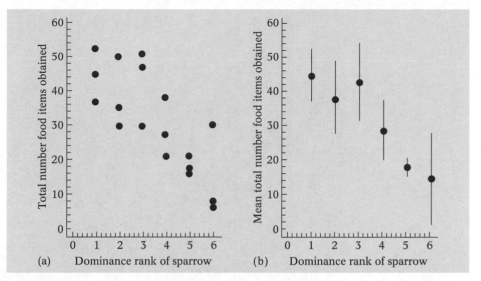

Figure 4.6 (a) The number of food items obtained during the period of observation by male house sparrows of different dominance status in groups of six (rank 1, most dominant; rank 6, least dominant, data for three groups at each rank). $r_s = -0.77$, $n = 18$, $p < 0.001$. (b) The mean number of food items obtained during the period of observation by male house sparrows of different dominance status in groups of six (rank 1, most dominant; rank 6, least dominant). $r_s = -0.77$, $n = 18$, $p < 0.001$. Bars represent standard errors.

Although there is a significant trend towards dominant males getting more food, the correlation is negative because we chose to use a rank of 1 for the most dominant male and a rank of 6 for the least dominant. Rankings are frequently ordered in this way, leading to the slightly odd situation of concluding a positive trend (e.g. dominants get more food) from what looks like a negative trend (the number of food items decreases with increasing rank number). There is no reason, of course, why dominance shouldn't be ranked the other way round (6 = most dominant, 1 = least dominant) so that a positive slope actually appears in the figure.

Sometimes when replicated observations are presented in a scattergram, they are presented as a single mean or median with appropriate standard error or confidence limit bars. Thus an alternative presentation of Fig. 4.6a is shown in Fig. 4.6b. Note that a different explanatory legend is now required because the figure contains different information.

In some cases, replication may not occur throughout the data set. Say we decided to sample a population of minnows (*Phoxinus phoxinus*) in a stream to see whether big fish tended to have more parasites. To avoid the difficulties of making accurate measurements of fish size in the field and possibly injuring the fish, we visually assess those we catch as belonging to one of six size classes. We then count the signs of parasitism on them and return them to the water. Because we have no control over the number of each size class we catch, we end up with more samples for some classes than for others. When we come to present the data, we could present them as individual data points for each size class (Fig. 4.7a) or condense replicated data for classes to means or medians (Fig. 4.7b).

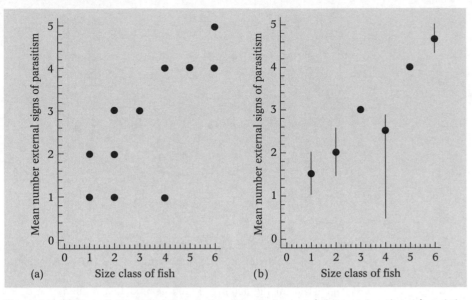

Figure 4.7 (a) The relationship between the size of minnows (arbitrary size classes) and the number of signs of parasitic infection observed on them. $r_s = 0.74$, $n = 11$, $p < 0.02$. (b) The relationship between the size of minnows (arbitrary size classes) and the mean number of signs of parasitic infection observed on them. $r_s = 0.74$, $n = 11$, $p < 0.02$. Bars represent standard errors.

In the latter case, only some points might have error or confidence limit bars attached to them because only some classes are replicated (*see* Fig. 4.7a). Data for those classes that are not replicated are still presented as single points.

It is, of course, important to remember that even where correlations are presented as mean or median values rather than independent data points, the correlation analysis itself (i.e. the calculation of the correlation coefficient) is still performed on the independent data points, not on the means or medians. Values of *n* are thus the same in Figs 4.6a and 4.6b and in Figs 4.7a and 4.7b. Correlations *can* be performed on summary statistic values, but obviously a lot of information is lost from the data and *n*-sizes are correspondingly smaller.

Presenting a regression analysis

Presenting a regression analysis is essentially similar to presenting a correlation except that a line needs to be fitted through the data points. If the trend isn't significant, so that a line should not be fitted, a figure probably isn't necessary in the first place. The details of calculating a regression line have been given earlier. You may sometimes come across regression plots that show confidence limits as curved lines above and below the regression line itself. However, we shall not be dealing with these here. For further information, see Sokal & Rohlf (1995).

As with correlation, data can be presented as independent points or, where replicated for particular *x*-values, as means or medians. Once again, where means or medians are presented, significance testing and the fitting of the line are still done using the individual data points, not the summary statistics. Figure 4.8 presents a regression of the effect of additional food during the breeding season

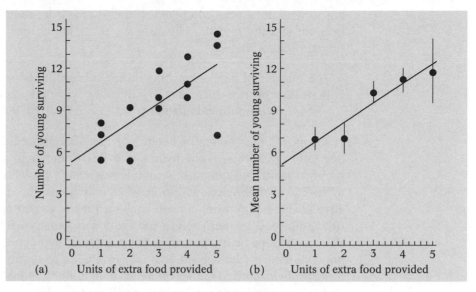

Figure 4.8 (a) The number of chicks surviving to their first winter in relation to the number of units of extra food provided during the breeding season in three populations of moorhen. $F = 13.27$, d.f. = 1,13, $p < 0.01$. (b) The mean number of chicks surviving to their first winter in relation to the number of units of extra food provided during the breeding season in three populations of moorhen. $F = 13.27$, d.f. = 1,13, $p < 0.01$. Bars represent standard errors.

on the number of young moorhens (*Gallinula chloropus*) surviving into their first winter in three study populations. Five different quantities of food were used and the three populations received them in a different order over a five-year experimental period. In Fig. 4.8a, the numbers for each population are presented separately; in Fig. 4.8b they are presented as means (± s.e.) across the three study populations.

4.2   Presenting results in the text

So far in this section, we have assumed that results will be presented as figures or tables. Figures and tables, however, take up a lot of space in a report and may not be justified if the result is relatively minor or there is a strict limit on the length of the report. In such cases, analyses can instead be summarised in parentheses in the text of the 'Results' section (*see later*). The usual form for a difference analysis is to quote the summary statistics, test statistic, sample size or degrees of freedom and *p*-value. Thus the information in Table 4.1a could easily be presented in the text as:

Treatment with gibberellin resulted in a significant increase in growth compared with non-treated controls (mean (± s.e.) height of treated plants = 14.75 ± 0.88 cm, $n = 12$; mean height of controls = 9.01 ± 0.63 cm, $n = 8$; $H = 12.22$, $p < 0.001$).

For a trend, it is usual to quote just the test statistic, sample size or degrees of freedom and the *p*-value. For instance, the information in Fig. 4.8a could be summarised as follows:

> The number of chicks hatching during a breeding season that survived into their first winter increased significantly with the amount of extra food provided within the population ($F = 13.27$, d.f. $= 1,13$, $p < 0.01$).

It is impossible to generalise about when an analysis could be presented in the text rather than in separate figures or tables. Sometimes, as we have said, it is simply a matter of limited space. However, rough guidelines might include the following: (a) difference analyses between only two or three groups; (b) corroborative analysis, supporting a main analysis already presented as a figure or table (for instance, if a main analysis showed a significant correlation between body size and fighting ability, a corroborative analysis might check that body size was not confounded with age and that the correlation could not have arisen because bigger individuals had more experience of fighting); (c) analyses providing background information (e.g. showing a significant sex difference in body size where this is germane to, say, an analysis of the diet preferences of the two sexes).

Units

Whether we are dealing with data in tables, figures or text, it is *essential* that appropriate units of measurement are included and cited with their conventional abbreviations (Box 4.1). Summary statistics are meaningless without them.

4.3 Writing reports

Just as figures and tables of data should be presented properly to ensure they are effective, so care must be taken in the text of a report. In the scientific community, reports of experiments and observations are usually published in the form of papers in professional journals, or sometimes as chapters in specialist books. In all cases, however, the aim is both to communicate the findings of a piece of research and provide the information necessary for someone else to repeat the work and check out the results for themselves. Both these elements are crucial and, as a result, scientific papers are usually refereed by other people in the same field to make sure they come up to scratch before being published. Not surprisingly, a more or less standard format for reports has emerged which divides the textual information into well-recognised sections that researchers expect to see and know how to refer to to find out about different aspects of the work. Learning to use this format properly is one of the most important goals of any basic scientific training. We shall therefore now discuss the general structure of a report and what should and should not go in each of its sections; then we shall develop a full report, incorporating our various points about text and data presentation, from some of our main example observations.

Box 4.1	Some common units of measurement and their conventional abbreviations

Length

Kilometre	km
Metre	m
Centimetre	cm
Millimetre	mm

Area

Square kilometre	km^2
Hectare	ha
Square metre	m^2
Square centimetre	cm^2
Square millimetre	mm^2

Volume

Cubic decimetre	dm^3
($\equiv$ litre)	(l)
Cubic centimetre	cm^3
($\equiv$ millilitre)	(ml)
Cubic millimetre	mm^3
($\equiv$ microlitre)	(μl)

Mass

Kilogram	kg
Gram	g
Milligram	mg
Microgram	mcg
Nanogram	ng

Time

Million years	My
Years	y
Hours	h
Minutes	min
Seconds	s

4.3.1 The sections of a report

There are five principal sections in a report of experimental or observational work: *Introduction*, *Methods*, *Results*, *Discussion* and *References*. Sometimes it is helpful to have some small additional sections such as *Abstract*, *Conclusions* and *Appendices*, but we shall deal with these later.

Introduction

The Introduction should set the scene for all that follows. Its principal objective is to set out: (a) the *background* to the study, which means any theoretical or previous experimental/observational work that led to the hypotheses under test. Background information is thus likely to include references to previously published work and sometimes a critical review of competing ideas or interpretations. It might also include discussion about the timing (seasonal, diel, etc.) of experiments/observations and, in the case of fieldwork, the reasons for choosing a particular study site; (b) a clear statement of the hypotheses and predictions that are being tested; and (c) the rationale of the study, i.e. how its design allows the specified predictions to be tested and alternatives to be excluded. The Introduction should thus give the reader a clear idea as to why the study was carried out and what it aimed to investigate. The following is a brief example:

> Reptiles are ectotherms and thus obtain most of the heat used to maintain body temperature from the external environment (e.g. Davies, 1979). Rattlesnakes (*Crotalus* spp.) do this by basking in the sun or seeking warm surfaces on which to lie (Bush, 1971). An increased incidence of snake bites in the state over the past two years has been attributed to a number of construction projects that have incidentally provided rattlesnakes with concrete or tarmac surfaces on which to bask (North, 2006). The aim of this investigation was to study the effect of the construction projects on basking patterns among rattlesnakes to see whether these might increase the exposure of people to snakes and thus their risk of being bitten. The study tests two hypotheses: (a) concrete and tarmac surfaces are preferred basking substrates for rattlesnakes and (b) such surfaces result in a higher than average density of snakes near humans.

When the reader moves on to the Methods and Results sections, they will then appreciate why things were done the way they were. Reading Methods or Results sections without adequate introductory information can be a frustrating and often fruitless business since the design of an experiment or observation usually makes sense only in the context of its rationale. As we shall see below, it is sometimes appropriate to include background material in the Discussion section, but in this case it should be to help develop an interpretation or conclusion, not an afterthought about information relevant to the investigation as a whole; if it is the latter, it should be in the Introduction.

Methods (or Materials & Methods)

The Methods section is perhaps the most straightforward. Nevertheless, there are some important points to bear in mind. Chief among them is providing enough detail for someone else to be able to repeat what you did *exactly*. Clearly, the precise detail in each case will depend on the investigation, but points that need attention are likely to include the following.

Experimental/observational organisms or preparations. The species, strain, number of individuals used, growth or housing conditions and husbandry, age and sex, etc. for organisms; the derivation and preparation and maintenance techniques, etc. for preparations (e.g. cell cultures, histological preparations, pathogen inoculations).

Specialised equipment. Details (make, model, relevant technical specifications, etc.) of any special equipment used. This usually means things like tape or video recorders, specialist computing equipment, spectrometers, oscilloscopes, automatic data-loggers, optical equipment such as telescopes, binoculars or specialised microscopes, centrifuges, respirometers, specially constructed equipment such as partitioned aquaria, choice chambers, etc. Run-of-the-mill laboratory equipment like glassware, balances, hotplates and so on don't usually require details, though the dimensions of things like aquaria or other containers used for observation and the running temperature of heating devices, etc. should be given. There should *not* be a section or list called 'equipment'. The Method section should not be written as a series of commands, but should be about what you did, and should be in the past tense.

Study site (field work). Where an investigation has taken place in the field, full details of the study site should normally be given. These should include its location (e.g. grid reference) and a description of its relevant features (e.g. size, habitat structure, use by people) and how these were used in the investigation. The date or time of year of the study may also be relevant.

Data collection. This should include details of all the important decisions that were made about collecting data. Again, it is impossible to generalise, but the following are likely to be important in many investigations: any pretreatment of material before experiments/observations (e.g. isolation of animals, drug treatment, surgical operations, preparation of cell cultures, staining); details of experimental/observational treatments *and controls*; sample sizes and replication; methods of measurement and timing; methods of recording (check sheets, tape recording, tally counters, etc.); duration and sequencing of experimental/observational periods; details of any computer software used in data collection. Of course, it is important not to go overboard. For instance, it isn't necessary to relate that a check sheet was ticked with a redball point pen rather than a black one, but if the pen was used to stimulate aggression in male sticklebacks (which often attack red objects) then it would be relevant to state that a ballpoint pen was used and that it was red.

Results

The Results section is in some ways the most difficult to get right. Many students regard it as little more than a dumping ground for all manner of summary and, worse, raw data. Explanation, where it exists at all in such cases, frequently consists of an introductory 'The results are shown in the following figures . . .' and a terminal 'Thus it can be seen . . .'. A glance at any paper in a journal will show that a Results section is much more than this. At the other extreme, explanation within the Results often drifts into speculative interpretation, which is more properly the province of the Discussion (*see below*).

A Results section should do two things and *only* two things: first, it should present the data (almost always in some summarised form, of course) necessary to answer the questions posed; and second, it should explain and justify the analytical approach taken so that the reasons for choice of test and modes of data presentation are clear. The section should thus include a substantial amount of explanatory text, but explanation should be geared solely to the analyses and presentation of data and not the interpretations or conclusions that might be inferred from them. An example might be as follows:

> Figure 1 shows that rattlesnakes are significantly more likely to be found on concrete (Fig. 1a) and tarmac (Fig. 1b) surfaces around dawn and dusk than around midday. Since many of the construction projects in the survey of snake bite incidence have involved highways (Greenbaum *et al.*, 1984), this temporal pattern of basking may result in highest snake/human encounter at times when public conveniences are closed and motorists are forced to relieve themselves at the roadside. Indeed Table 1 shows a strong association for three highways between time of day and number of motorists stopping by the roadside.

It is also important that all the analyses and presentations of data involved in the report appear in the Results section (as figures, tables or in the text) and only in the Results section; no analysis should appear in any other section.

Discussion

The Discussion is the place to comment on whether the results support or refute the hypotheses under test and how they relate to the findings of other studies. The Discussion thus involves interpretation and reasonable speculation, with further details about the material investigated and any corroborative/contradictory/background information as appropriate. As we have said, however, while the Discussion may flesh out, comment, compare and conclude, it should not bring in new analysis. Neither should it develop background information that is more appropriate to the Introduction (*see earlier*). The kind of thing we'd expect might be as follows:

> The results suggest that concrete and tarmac surfaces are not favoured for basking by rattlesnakes in comparison with broadly equivalent natural surfaces when relative area is taken into account. One reason for this might be the greater proximity and greater density of cover close to the natural surfaces sampled. Many snakes (Jones, 1981), including rattlesnakes (Wilson, 1998), prefer basking areas within a short escape distance of thick cover. Despite not being preferred by snakes, the greater incidence of bites on concrete and tarmac surfaces can be explained in terms of the greater intensity of use of these surfaces by humans. However, Wilson (1998) has noted that the probability of attack when a snake is encountered increases significantly if there is little surrounding cover. The paucity of cover around the concrete and tarmac samples may thus add to the risk of attack in these environments.

References

Your report should be referenced fully throughout, with references listed chronologically in the text and alphabetically in a headed References section at the end.

References styles vary enormously between different kinds of report so there is no one accepted format. However, a style used very widely is illustrated below and we suggest using it except where you are explicitly asked to adopt a different style. In this style, references in the text should take the form:

> . . . Smith (1979, 1980) and Grant *et al.* (1989) claim that, during a storm, a tree 10 m in height can break wind for over 100 m (*but see* Jones & Green, 2002; Nidley, 1999, 2001) . . .

In the References list at the end, journal references take the form:

> Grant, A. J., Wormhole, P. & Pigwhistle, E. G. (1989) Tree lines and the control of soil erosion. *Int. J. Arbor.* **121**, 42–78.
> Jones, A. B. & Green, C. D. (2002) Soil erosion: a critical review of the effect of tree lines. *J. Plant Ecol.* **97**, 101–107.
> Smith, E. F. (1979) Planting density and canopy size among deciduous trees. *Arbor. Ecol.* **19**, 27–50.
> Smith, E. F. (1980) Planting density and growth rate among deciduous trees. *Arbor. Ecol.* **20**, 38–52.

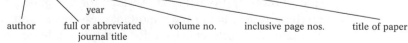

author   full or abbreviated journal title   year   volume no.   inclusive page nos.   title of paper

for books they take the form:

> Nidley, R. (1999) *Deforestation and its impact on national economies.* Hacker Press, London.

publisher   place of publication   title of book in italics and for chapters in edited volumes the form:

> Nidley, R. (2001) Economic growth and deforestation. In *Sustainable economics and world resources*, eds A. B. Jones & C. D. Green, pp. 64–78. Hacker Press, London.

Where more than one source by a particular author (or set of authors) in a particular year is referred to, the sources can be distinguished by using lower case letter suffixes, e.g. (Smith, 1976a, b) indicates that you are referring to two reports by Smith in the year 1976. The order in which you attribute a, b, c, etc. is determined by the order in which you happen to refer to the publications in your report, not the order in which they were published in the relevant year.

Personal observations and personal communications. Although most of the references you will make will be to work by other people, or yourself, that has been published in some form, it is occasionally appropriate to refer to unpublished observations. This usually arises where some previous, but unpublished, observation is germane to an assumption, fact, technique, etc. that you are relying on in your own report. If such observations are your own, they can be referred to in the text as '(personal observation)' or '(pers. obs.)'. If they have been reported to you by someone else, then they can be referred to as, for example, '(P. Smith, personal communication)' or '(P. Smith, pers. comm.)' – note that the name of the person providing the information is given as well.

Abstract. A small, but important, section of many scientific reports, certainly published ones like papers in learned journals, is the *Abstract*. This is a short (often strictly word-limited) summary of the aims and main findings of the investigation. The idea is to provide the reader with a quick overview of what was done and what was interesting about it, so that the reader can decide whether they want to read the report in more detail. Abstracts are particularly important in the case of published reports because they are often made available online to people browsing the various searchable scientific literature databases (*see* Box 1.1). They are thus a useful 'shop window' for available studies on the chosen topic. Increasingly, as part of the general push for greater public awareness of science, abstracts are now also being made available in the form of 'lay summaries', meaning that they are redrafted in simple, everyday language that people without formal scientific training can understand (*see* section 4.4 below); often they are sent out to the media. Whether or not you intend to try to publish your report, however, producing an Abstract for it is good practice because it makes you think clearly about the important messages in your work and express them succinctly. An Abstract is included in the example report in Box 4.2 to illustrate the point.

Other sections of a report
In some cases, there may be additional sections to a report.

Conclusions. Sometimes, especially where analyses and interpretations are long and involved, it is helpful to highlight the main conclusions in a tail-end section so that the reader finishes with a reminder of the 'take-home' message of the investigation. In general, however, the Abstract serves the function of a conclusion, and therefore you have either an Abstract or a Conclusion, but not both.

Appendix. Occasionally, certain kinds of information may be incorporated into an Appendix. Such information might include the details of mathematical models or calculations, detailed background arguments, selective raw data or other aspects of the study that potentially might be of importance to readers but which would clutter up and disrupt the main report were they to be included there. Appendices are thus for informative asides that might help some readers but perhaps distract others. It follows, therefore, that appendices should be used selectively, sparingly and for a clear purpose, not as a dumping ground for odds and ends on the grounds that they might just turn out to be useful.

Use of abbreviations. It is also worth saying something about the use of abbreviations. Many long-winded technical and jargon terms are often abbreviated in reports, papers and books. This is common practice and perfectly acceptable, as long as abbreviations are defined at their first point of use and conventions are adhered to where they exist (some acronyms, for example, are so well established that people are hard put to recall the full terminology). Thus:

> The high vocal centre (HVC) in the forebrain of birds is associated with the production of song. The volume of the HVC also varies with the complexity of song in different species.

and

> To see whether there was any effect of site on the frequency of calling, we carried out a one-way analysis of variance (ANOVA). The results of the ANOVA suggested that site had a profound effect.

present no problem, whereas:

> The HVC in the forebrain of birds is associated with the production of song.

or

> To see whether there was any effect on the frequency of calling, we carried out an ANOVA.

leaves the uninitiated little the wiser.

While abbreviations and acronyms are acceptable, however, they should be used judiciously. Littering text with them is a sure way to destroy its readability and confuse the reader.

4.3.2 Example of a report

Having outlined the general principles of structuring a report, we can finish off by illustrating them more fully in a complete report (see Box 4.2). The report is

| Box 4.2 | Example report |

The effect of body size on the escalation of aggressive encounters between male field crickets (*Gryllus bimaculatus*)

Abstract

Fighting is a costly activity; it takes time and energy and risks injury or even death. One way animals may be able to reduce the cost of aggressive competition is by assessing their chances of winning before becoming involved in a fight. Various attributes of opponents might provide information about the likelihood of winning, an obvious one being body size. When male field crickets (*Gryllus bimaculatus*) were allowed to interact in a sand-filled arena, encounters were more likely to become aggressive as the difference in body size between opponents declined, suggesting relative body size was important in assessing whether or not to escalate into a fight. However, the results also suggested that experience of winning or losing a fight itself affected the tendency to initiate and win subsequent fights. Aggressive encounters between male crickets may thus depend on both assessment of opponents and the degree of confidence of competing individuals at the time of encounter.

Introduction

Fighting is likely to be costly in terms of time and energy expenditure and risk of injury to the individuals involved. We might thus expect natural selection to have favoured mechanisms for reducing the likelihood of costly fights. One way animals could reduce the chance of becoming involved in an escalated fight is to assess their chances of winning or losing against a given opponent before the encounter escalates into all-out fighting. There is now a substantial body of theory (e.g. Parker, 1974; Maynard Smith and Parker, 1976; Enquist *et al.*, 1985) suggesting how assessment mechanisms might evolve and much empirical evidence that animals assess each other during aggressive encounters (e.g. Davies & Halliday, 1978; Clutton-Brock *et al.*, 1979; Austad, 1983). Since the outcome of a fight is likely to be determined by some kind of difference in physical superiority between opponents, features relating to physical superiority might be expected to form the basis for assessment.

Male field crickets compete aggressively for ownership of shelters and access to females (*see* Simmons, 1986). Casual observation of male crickets in a sand-filled arena suggested that body size might be an important determinant of success in fights, with larger males winning more often (pers. obs.). This is borne out by Simmons (1986), who found a similar effect of body size in male *G. bimaculatus*. Observations also showed that aggressive interactions progressed through a well-defined series of escalating stages (*see also* Simmons, 1986, and e.g. Brown *et al.*, 2006 and Nosil, 2002 for other cricket species) before a fight ensued. One possibility, therefore, is that these escalating stages reflect the acquisition of information about relative body size, and interactions progress to the later, more aggressive, stages only when opponents are closely matched in size and the outcome is difficult to predict. This study therefore tests two predictions arising from this hypothesis:

1. large size will confer an advantage in aggressive interactions among male crickets, and
2. interactions will escalate further when opponents are more closely matched in size.

Methods

Four groups of six virgin male crickets were used in the experiment. All males were derived from separate, unrelated stock colonies a week after adult eclosion so each group comprised arbitrarily selected, unfamiliar males on establishment. Crickets were maintained on a 12 h : 12 h light : dark cycle that was shifted by 4 h to allow observation at periods of peak activity (Simmons, 1986). Before establishing a group, the width of each male's pronotum (thorax) was measured at its widest point using Vernier calipers and recorded as an index of the male's body size (the pronotum was chosen because it consists of relatively inflexible cuticle that is unlikely to vary between observations or with handling; adult body size is determined at eclosion so does not change with age). The dorsal surface of the pronotum of each male was then marked with a small spot of coloured enamel paint to allow the observer to identify individuals.

Groups were established in glass arenas ($60 \times 60 \times 30$ cm^3) with 2-cm deep silver sand substrate. Each arena was provided with water-soaked cotton wool in a Petri dish and two to three rodent pellets. No shelters or other defendable objects were provided to avoid bias in the outcome of interactions due to positional advantages. Arenas were maintained under even 60 W white illumination in an ambient room temperature of 25 °C throughout the experiment.

The six males in a group were introduced into their arena simultaneously and allowed to settle for 5 min. They were then observed for 30 min, during which time all encounters between males were dictated onto magnetic tape, noting: (a) the individuals involved, (b) the individual initiating the encounter (the first to perform any of the components of aggressive behaviour – *see below*), (c) the individual that won (decided when one opponent first attempted to retreat) and (d) the components of aggressive behaviour used by each opponent during the encounter. Following Simmons (1986), the aggressive behaviours recognised here are shown in Table 1.

Results

Do larger males tend to win aggressive encounters? To see whether larger males tended to win more often, the percentage of encounters won by each male in the four groups was

Table 1 Degree of escalation increases from Aggressive stridulation to Flip. Each behaviour can thus be ascribed a rank escalation value ranging from 1 (low escalation) to 6 (high escalation).

Behaviour	Description	Escalation ranking
Aggressive stridulation	One or both males stridulate aggressively. This may occur on its own or in conjunction with other aggressive behaviours	1
Antennal lashing	One male whips his opponent with his antennae	2
Mandible spreading	One male spreads his mandibles and displays them to his opponent	3
Lunge	A male rears up and pushes forward, butting the opponent and pushing him backwards	4
Grapple	Males lock mandibles and wrestle	5
Flip	One male throws his opponent aside or onto his back. Re-engagement was rare following a Flip	6

plotted against pronotum width (Fig. 1). A significant positive trend emerged. Figure 1, however, combined data from all four groups. Did the relationship hold for each group separately? Spearman rank correlation showed a significant relationship in three of the four groups ($r_s = 0.94$, 0.99, 0.97 ($p < 0.05$ in all cases) and 0.66 (ns), $n = 6$ in all groups, one-tailed test).

If there is a size advantage as suggested by Fig. 1, we might expect larger males to initiate more encounters than smaller males since they have more to gain. Figure 2 shows a significant positive correlation between pronotum width and the percentage of the recorded encounters for each male that was initiated (*see* Methods) by that male. As expected, therefore, larger males tended to be the initiator in more of their encounters.

One possibility that arises from Figs 1 and 2 is that the apparent effect of body size was an incidental consequence of the tendency to initiate. There may be an advantage to initiating itself, perhaps because an individual initiates only when its opponent's ability to retaliate is compromised (e.g. it is facing away from its attacker). If the males doing most of the initiating in the groups just happened to be the bigger ones, the initiation could underlie the apparent effect of body size on the chances of winning. To test this, the percentage encounters won by each male when he was the initiator was compared with the percentage won when he was not. The

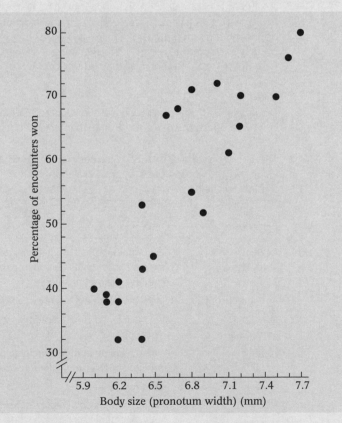

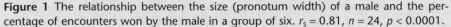

Figure 1 The relationship between the size (pronotum width) of a male and the percentage of encounters won by the male in a group of six. $r_s = 0.81$, $n = 24$, $p < 0.0001$.

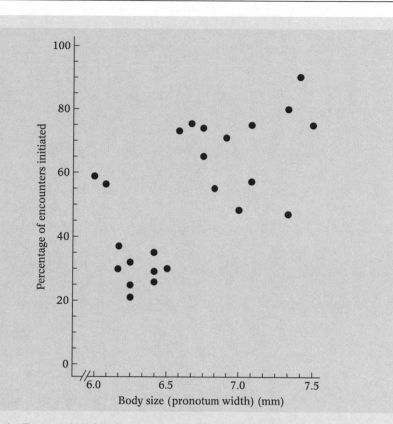

Figure 2 As Figure 1 but for the relationship between male size and the percentage number of encounters initiated by the male. $r_s = 0.63$, $n = 24$, $p < 0.03$.

analysis showed no significant difference ($U = 82$, n_1, $n_2 = 19$, ns[†]) between the two conditions.

Does difference in body size affect the degree of escalation in encounters? Figure 3 shows the relationship between the ratio of pronotum width for pairs of opponents and the degree of escalation of their encounters. A ratio of one indicates equal size, and ratios greater than one increasing departure from equality. Degree of escalation is measured as the maximum rank value (1–6, *see* Methods, Table 1) recorded during an encounter. As predicted, the figure shows a significant negative correlation between size ratio and degree of escalation so that escalated encounters were more likely between opponents that were closely matched in size. The trend in Fig. 3 is for data across all groups. Does the trend hold within individual males? Correlation analysis for those males (four) that were involved in five or more encounters with opponents of different relative size suggests that it did, although the trends were significant in only two cases ($r_s = -0.96$, $n = 7$, $p < 0.05$; $r_s = -0.72$, $n = 5$, ns; $r_s = -0.76$, $n = 6$, ns; $r_s = -0.99$, $n = 6$, $p < 0.05$, one-tailed

---

[†] Although it is perfectly legitimate to use a Mann–Whitney *U*-test here, the fact that we are actually comparing data for the two conditions (initiated versus non-initiated encounters) *within* males means we could have used a different sort of two-group difference test (e.g. a Wilcoxon matched-pairs signed ranks test) which takes this into account.

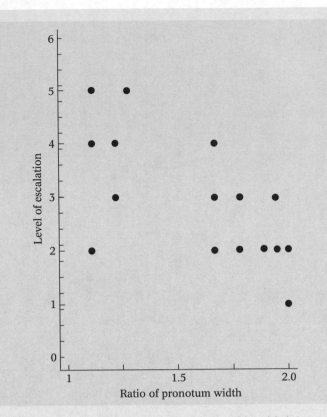

Figure 3 The relationship between ratio of pronotum widths of fighting males and the maximum level of escalation (1–6, *see* Methods) reached in fights. $r_s = -0.71$, $n = 20$, $p < 0.002$.

test). A similar analysis of the relationship between size ratio and the duration of aggressive encounters also showed a significant tendency for closely matched males to fight for longer ($r_s = 0.62$, $n = 20$, $p < 0.002$).

Does the experience of winning or losing affect subsequent interactions? The results so far are consistent with males assessing each other on the basis of relative size. However, it is possible that the experience of winning or losing a fight might itself influence a male's approach to subsequent encounters. A male that has just won a fight, for example, might assess his chances of winning the next one as being higher than if he had just lost (a 'confidence' effect). To see whether this was the case, the outcome of fights for males that were recorded as having ten or more encounters during the observation period were analysed according to whether the male won or lost his first recorded encounter. A Mann–Whitney U-test showed that males winning their first encounter won a significantly greater proportion of their subsequent encounters than those losing their first encounter ($U = 8$, $n_1 = 6$, $n_2 = 9$, $p < 0.05$). They also initiated a significantly greater proportion of the encounters ($U = 9$, $n_1 = 6$, $n_2 = 9$, $p < 0.05$). Interestingly, there was no significant difference in body size between males winning versus losing their first encounter ($U = 16$, $n_1 = 6$, $n_2 = 9$, ns), so the 'confidence' effect seemed to be to that extent independent of the effect of size.

Discussion

The results bore out both predictions about the effects of body size on the outcome of aggressive interactions between male field crickets: larger males were more likely to win and escalation was more likely between closely matched opponents. This is consistent with the outcome of fights being largely a matter of physical superiority and with the structuring of interactions into a well-defined series of escalating stages reflecting assessment.

The fact that larger males were more likely to initiate an interaction could mean that the relative size of a potential opponent is assessable in advance of physical interaction. However, it could also reflect a general confidence effect arising from previous wins by larger males (males may initiate according to the simple decision rule 'if I won in the past, I'll probably win this time, so it is worth initiating'). Indeed, Simmons (1986) presents evidence that the number of past wins has a positive influence on the tendency for males to initiate, a result consistent with the apparent confidence effect in the present study (but see e.g. Brown *et al.*, 2006 and Nosil, 2002 for more equivocal results from other species). Alternatively, initiation could reflect individual recognition, with males picking on those individuals against whom they have won in the past. Since this study did not record encounters independently of the performance of one of the categories of aggressive behaviour, it is not possible to say whether initiations against particular opponents occurred more or less often than expected by chance. Whatever the basis for deciding to initiate, however, there was no evidence that initiation itself conferred an advantage in terms of the outcome.

Although no resources (shelters and females) were available in the arenas, the size advantage in the aggressive interactions recorded here is in keeping with the tendency for larger males to take over shelters and mate successfully with females (Simmons, 1986). While females prefer to mate with males in or near shelters (because these provide good oviposition sites and protection from predators), they will mate with males encountered in open areas (Simmons, 1986). Aggression between males in the absence of shelters or females may thus reflect an advantage to reducing competition should a female happen to be encountered.

References

Austad, S. N. (1983) A game theoretical interpretation of male combat in the bowl and doily spider, *Frontinella pyramitela*. *Anim. Behav.* **19**, 59–73.

Brown, W. D., Smith, A. T., Moshalik, B. & Gabriel, J. (2006) Aggressive contests in house crickets: size, motivation and information content of aggressive songs. *Anim. Behav.* **72**, 225–233.

Clutton-Brock, T. H., Albon, S. D., Gibson, R. M. & Guinness, F. E. (1979) The logical stag: adaptive aspects of fighting in red deer (*Cervus elaphus* L). *Anim. Behav.* **27**, 211–225.

Davies, N. B. & Halliday, T. R. (1978) Deep croaks and fighting assessment in toads, *Bufo bufo. Nature* **274**, 683–685.

Enquist, M., Plane, E. & Roed, J. (1985) Aggressive communication in fulmars (*Fulmarus glacialis*) competing for food. *Anim. Behav.* **33**, 1107–1120.

Maynard Smith, J. & Parker, G. A. (1976) The logic of asymmetric contests. *Anim. Behav.* **24**, 159–175.

Nosil, P. (2002) Food fights in house crickets, *Acheta domesticus*, and the effects of body size and hunger level. *Can. J. Zool.* **80**, 409–417.

Parker, G. A. (1974) Assessment strategy and the evolution of animal conflicts. *J. Theor. Biol.* **47**, 223–243.

Simmons, L. W. (1986) Inter-male competition and mating success in the field cricket, *Gryllus bimaculatus* (de Geer). *Anim. Behav.* **34**, 567–579.

one that might arise from some of the experiments we proposed earlier in the main examples, in this case aggression in crickets.

4.4   Writing for a more general readership

So far, we've looked at writing reports on the assumption that the target readership is other biologists, whether these are lecturers or tutors in an educational setting, or fellow researchers. The key concerns with this kind of writing are: (a) adhering to the appropriate conventions of scientific reporting, and (b) the technical clarity with which the hypotheses and predictions under test, the manner in which the study was carried out, the statistical analyses and the background and conclusions to the study are presented for critical professional scrutiny. This can make for a rather dry and jargon-strewn read for anyone coming to it out of general interest, and perhaps with little or no formal scientific education (the proverbial 'intelligent layman'). Nevertheless, there are many good reasons why such a person should be able to appreciate what has been done, and why, in scientific research (not least because they are probably helping to pay for it out of their taxes!), and equally good reasons why scientists themselves should make the effort to render their work accessible rather than leaving it to other people, such as journalists, who often have a poor grasp of the work or try to sensationalise it into a newsworthy story.

In the UK, the art of explaining science to the 'intelligent layman' now has its own buzz phrase, the 'Public Understanding of Science' (or PUS, to use its slightly unfortunate acronym), one enthusiastically promoted by government and honoured with an eponymous chair at Oxford. In its wake, scientists are actively encouraged to engage with the media to promote and explain their work, and many universities and colleges offer courses on popular scientific writing and other public communication skills. Not surprisingly, therefore, science students increasingly see career opportunities in the wider communication of their subject. So what exactly does 'wider communication' entail? Well, of course, it depends to some extent on what you're trying to talk about and to whom, but there are some general points that will help keep you on the right track.

Avoid jargon

Probably the first golden rule is to avoid, as far as possible, using any technical jargon or unfamiliar scientific terminology. For example, the following passage in a scientific article on immune response to infection:

In a study of European bank voles (*Myodes glareolus*) high burden populations were characterised by high levels of plasma corticosterone and testosterone and high fluctuating asymmetry (FA) in hind foot length (a putative measure of developmental instability; e.g. Palmer & Strobeck, 1986; Møller & Swaddle, 1997), but reduced aggressiveness among males (Barnard *et al.*, 2002, in review).

might be rendered as:

> Bank voles are small woodland rodents, rather like mice, but with shorter ears and tails. They are common across Europe where, like other wild mammals, they are often infected with various parasites, such as tapeworms and fleas. In areas where infections are very severe, voles often show evidence of stress and poor physical development, and males are less aggressive than their counterparts elsewhere.

This has managed to recast the whole thing in easily understood, everyday language that most people would be able to follow. It tells them what voles are (many may not know), and what the piece is talking about when it refers to parasites. It also leaves out all the clutter of scientific referencing, essential for the professional scientist, but a distraction for the general reader.

Sometimes, however, the use of a certain amount of technical jargon may be inescapable, or even helpful in giving some idea of the scientific approach behind a report. If so, *explain it* in simple everyday terms. For instance, a different version of the passage above might have gone as follows:

> Voles from areas with very high levels of parasite infection show poor physical development. This can be judged by something scientists call 'fluctuating asymmetry', which is a measure of the difference in size between the same parts of the body, say the front or hind legs, on the left and right sides. The bigger the difference, the greater the hardship the animal is assumed to have experienced during its development.

This includes the technical term 'fluctuating asymmetry', because the reader might wonder exactly how poor physical development was measured, but immediately counters its potential to confuse with a simple description of what it means.

Use catchy analogies

Most people scanning a newspaper or magazine for articles that might interest them need something to draw their attention to a particular piece in the first place, and then to maintain their attention as they read it. A snappy title might achieve the first aim, but ensuring the second can be a bit trickier. A favourite device is to draw parallels with familiar aspects of our own experience. The old adage that 'sex sells newspapers', for example, is as true now as it ever was, and anything that reflects on, or can be compared with, our own sexual behaviour, however tangentially, can usually be assured of at least a passing glance. Health, war, culture and intelligence are other good 'hooks' for getting attention, because they touch on our preoccupations or views on what makes us uniquely human. Some popular science writers use characters or events from literature or works of art to draw analogies and create striking imagery, Stephen Jay Gould being an enthusiast of the style. Indeed, cleverly appropriate metaphors are an effective way in general to get points and ideas across, as illustrated by Richard Dawkins's famous book *The Selfish Gene* (first published in 1976), a successful popular account of the principle of natural selection that uses the 'selfish' metaphor to characterise the effects of selection on how organisms behave. As authors like Dawkins have discovered, however, it is often wise to emphasise that you are using metaphors as just that, and not in a literal sense (for instance, the 'selfish' gene metaphor simply captures the idea that genes code for characteristics that enhance their chances of being passed on to the next generation; it doesn't imply they are thoughtlessly self-centred in the sense we might use 'selfish' to describe other people). Bearing these points in mind, we might lead into our vole story with something like the following:

> The stressed office worker falling prey to every passing infection is a familiar cliché of our pressured, industrialised lives, and good medical research suggests it has a sound basis in fact. But is it something unique to us and our artificially hectic lifestyle? Evidence from the leafy glades of Europe's woodlands suggests not. A recent study of voles in the Polish forests of Mazury has revealed an association between stress and infection with parasites that has more than an echo of sick office worker syndrome.

So here we've used a familiar everyday analogy to get the reader's attention and cue them in to the message of what follows; the reader knows to expect something about how stress relates to disease and is thus primed to follow the piece. Box 4.3 presents a general readership piece along these lines on the cricket study reported in formal scientific style in Box 4.2.

4.5 Presenting in person: spoken papers and poster presentations

The outcomes and conclusions of your research can also be presented in person in the form of a talk, or in a fashion somewhat intermediate between a written and spoken presentation known as a poster presentation. As with the written word, the audience for a talk or poster can be diverse – from a group of fellow students reporting their research at the end of a field course, to the serried ranks of a thousand or so delegates at an international conference. The major difference from the traditional written word is that you are presenting the information live (poster presentations usually involve sessions where you explain and defend your findings personally to interested readers), and this difference can be enough to test the composure of even the most confident individual. At one level there are as many styles of spoken and poster presentations as there are presenters, and, to a lesser extent, each person in the audience will have a slightly different reaction to any given style. However, there are features that are shared by good presentations, and we discuss some of these here.

4.5.1 A presentation is not a written report

As we explained in section 4.4, writing for a general readership needs a different approach from writing a scientific paper. The same is true of spoken and poster presentations; indeed, they have more in common with writing for a general audience than writing a paper. And while presenting a spoken paper is sometimes

Box 4.3 **A general readership account of fighting in crickets (from Box 4.2)**

CANNY CRICKETS SIZE UP THE OPPOSITION

As any military commander worth his salt will tell you, 'know thine enemy' is a wise rule when it comes to a showdown against a powerful opponent. Rush in on impulse and it probably won't take long to regret it. This turns out to be as true for other animals as it is for ourselves. And, while they may not be hot on handy aphorisms, our fellow planetary inhabitants can be as wily as any military strategist when it comes to fisticuffs. Take the humble field cricket (*Gryllus bimaculatus*), for example.

Field crickets are distributed widely throughout southern Europe, where males contribute to the ringing chorus of insects that fills the evening air in summer. The reason males put all this effort into calling, however, is not to provide musical accompaniment for human bystanders, but to set about the serious business of competing for females. And here, the males' tuneful performance is just for starters.

As well as providing a sound beacon by which females can home in on a male, calls are picked up by other, rival, males. Since calling is costly – it takes time and effort, and can attract predators – these males can avoid paying the cost themselves by keeping quiet and intercepting the females attracted by their rival's efforts. Thus, callers can rapidly find themselves having to compete for their hard-won female with a bunch of freeloaders. Calling aggressively at them may have some effect, but often not much, so it's not long before things start to get physical. This is where careful choices have to be made.

Observations of crickets in the field, and under laboratory conditions mimicking those in the field, show that males don't just get stuck in indiscriminately. Interactions follow a well-defined sequence of behaviours that gradually escalate in their level of aggressiveness. If an opponent doesn't give way when lashed with the aggressor's antennae, for instance, the encounter might progress to locking jaws and wrestling. Ultimately, the larger of the contesting males tends to win, but the way encounters progressively up the ante strongly suggests males are weighing each other up before going all out. If so, an obvious prediction is that fights should last longer and be more intense as the difference in size between opponents decreases, because it would be harder for each to tell who was the biggest. Some recent experiments by scientists at the University of Nottingham, UK have shown that this is exactly what happens.

When pairs of males were carefully measured and then allowed to compete with each other, they were much more aggressive, and fought for longer, when closely matched in size. Thus, if reliable information about relative size and strength was not forthcoming when opponents prospectively probed each other, there appeared to be nothing for it but to set to. But things weren't quite as simple as this. While assessing the opposition certainly seemed to be part of the process, it turned out that males were also affected by their previous experience of winning or losing. If they had just won a fight, they were more likely to enter the next one with gusto and win that too. If they'd lost, however, they were more circumspect the next time and tended to lose again. This suggests that the degree of confidence with which a male approaches a fight has an important role to play, and that, in the world of crickets, just as in our own, he who hesitates is lost.

termed *reading a paper*, the result, if taken literally, usually makes for deadly listening for the simple reason that written and spoken language are generally very different in style. Written prose is usually more formal in structure and use of language than the spoken word, which abounds with illustrative figures of speech, amusing asides and informal phraseology that is easy on the ear. For instance, in a published paper we might write:

> In summary, therefore: (a) female thargs are on average 50 per cent larger than males, (b) only males over 4 kg in body weight initiate courtship, and (c) only males over 5 kg mate successfully.

But we'd probably *say* something more like:

> OK, so what can we conclude from all this? Well for one thing it's clear that females are the larger sex in thargs, which may mean that only relatively big males stand a chance of success in courtship. This may be why it's only 4 kg plus males that attempt courtship and only 5 kg plus males that actually get anywhere with it.

Very few people can write successfully in spoken language, so a paper read out verbatim usually sounds stilted and wooden. One past master of the art of writing in spoken English (in his case) is the evolutionary biologist Richard Dawkins, who customarily reads his talks from a prepared script, but he is a real exception.

The same is really true of poster presentations. One thing a poster absolutely should *not* be is a written paper pinned to a board, however attractively the author may feel the pages are set out. The art of good poster presentation owes more to the skills of the advertising agency than of the professional writer: it's all about getting a message over clearly and immediately (people browsing anything from 20–30 upwards to 500 or more posters are not going to hang around reading sheets of dense prose; they want headlines and images they can take in at a glance). We have occasionally seen a scientific paper stapled to a poster board at a conference, but this is more likely a consequence of a catastrophe befalling the original, intended poster than a serious attempt to use a written report as a poster presentation.

4.5.2 General considerations

Know your audience

It should probably go without saying that the content of a presentation will depend on the audience. A more general audience, such as a local natural history society or schoolchildren, will require more background to the study and less technically demanding language than a group of specialists, such as we should find in a research group. As a rule, the more general the audience, the more a presentation should deal in simple (but nonetheless accurate) take-home messages rather than the detailed mechanics of how the messages were arrived at (*see below*). We are not advocating skipping important qualifying information in general presentations, just recommending that care is taken not to lose the interesting facts and findings in it all. The key is therefore that the pitch and style need to be geared appropriately for different audiences.

Know your allocation

As well as choosing an appropriate style for your presentation, there is almost always another challenge to be met: it has to fit into a predetermined allocation of time or space. Where talks at conferences are concerned, this can be extremely short, 10 minutes or so sometimes, but may be up to 50 minutes or an hour if you're a key speaker. People often underestimate how long it takes to speak to a slide in a talk, and therefore tend to have too many slides for the time. As a general rule you should allow about two minutes per slide, which means around 10 slides for a 20-minute talk. Remember to talk the audience through each slide, pointing to the relevant pieces of information to which you are referring; don't just put the slide up and talk as if it wasn't there. Poster sizes are usually determined by the size and shape of available boards, and, of course, the number of posters that have to be squeezed onto them. Either way, it is an essential skill of giving spoken or poster presentations that you can tailor them to a required slot. It is thus vital to know the time allotted for your spoken presentation or the dimensions of the poster space. If in doubt (though you shouldn't be), err on the short side – few in the audience will mind if the talk finishes a few minutes early or if your poster doesn't completely fill the poster board. However, taking more than your allocation of time or space will usually elicit negative reactions, or result in you being cut off mid-presentation by an irritated session chairperson or not being able to put your poster up at all.

Know your aim

A presentation is sometimes seen as an opportunity (or temptation) to report every last detail of your study. This is a fatal mistake! There is never enough time or space to do it, and nobody's attention span would take it even if there was. Carefully, but ruthlessly, choose the aspects of your study that will be of most interest to the audience. Thus, if you are reporting interesting mating behaviour in your study species, focus on what makes it interesting and perhaps sets it apart from mating behaviour in other, similar species rather than on exactly how many dreary wet days you spent freezing in a remote hide collecting the data, or how you finally arrived at the clever statistical analysis you used. Make sure also that your points can be readily conveyed in the presentation – a personal demonstration of your study species' mating display on a table top could be the highlight of a spoken presentation, but you will need to think of another way of conveying the information if presenting a mostly unattended (by you) poster – a carefully chosen, clear photograph being an obvious option.

Less is more

Although a cliché, the spirit of this little aphorism has probably contributed more to good presentations than almost anything else. Whether the text will appear as a slide in a talk or as a panel on a poster, it is often surprising just how few words are needed to convey the meaning. For example, a first attempt at an introductory slide for a talk about fighting in crickets might set the agenda as follows:

Introduction

- Fighting is likely to be expensive in terms of time and energy and the risk of injury to individuals
- We might thus expect selection to have favoured mechanisms to reduce the likelihood of costly fights
- One way this could be achieved is by individuals assessing their chances of winning against an opponent before getting involved in a fight
- There is now a substantial body of theory suggesting how assessment mechanisms might evolve, and much evidence that animals do assess each other during aggressive encounters
- Male field crickets (*Gryllus bimaculatus*) compete aggressively for ownership of shelters and access to females, and casual observation suggests that body size might be an important determinant of success in fights

---

However, this is a lot of text, which will probably take more time than the slide is on the screen to read properly. The point is, you don't need to use full sentences to convey the essential message, and less text means you can use larger fonts and maybe some associated images to make the whole thing more digestible. The following box might be an effective distillation of the full-blown text in the box above.

INTRODUCTION

- Fighting can be costly
- Selection may thus favour reduced fighting
- Perhaps through assessing opponents first
- We can model the evolution of assessment
- And test models using fighting in crickets

This style is often referred to as *telegraphese*, because messages sent by telegram were charged by the word and shorter messages saved money. (An extreme example was an exchange of telegrams between the nineteenth-century French author Victor Hugo and his publisher: Hugo sent a telegram containing only a question mark, correctly assuming his publisher would understand he was asking how the sales of his recent novel were going – the publisher replied with an exclamation mark, indicating they were extremely good).

Less is also more when it comes to images (like photographs or video windows) and embellishments (like borders, cartoons and coloured text boxes), so the advice is the same as for words: unless an image or embellishment is really central to the message you are trying to convey, or essential for clarity, carefully consider whether it merits inclusion (*see also* Box 4.4 *below*). The aim is to design the presentation around the core message, not to produce a kaleidoscope of information for its own sake.

Images versus text

'A picture is worth a thousand words' is another cliché, but again captures the spirit of good presentation. Images and graphics can often replace text or they can enhance and support it, perhaps by illustrating the environment in which the study took place, or the lifestyle of the species concerned. For instance, the detailed list of escalating aggressive behaviours in Table 1 of the report in Box 4.2 could be replaced by a series of three to four photographs of the relevant behaviours with simple accompanying labels. A picture or two of fighting crickets might also helpfully embellish some of the introductory, data or summary slides that would comprise a talk, or figure as background embellishments in a poster.

Tables of data published in papers (for example, like that in Table 4.3) can very rarely be included in presentations as they are. Apart from the difficulty of reading the detail once the table becomes larger than a 3×3 matrix, the information is not in the best form for the audience to appreciate the interesting differences and similarities in the data. Converting tables to graphs, such as bar charts or means plots, helps to make such differences clear (*see* Figs 4.1, 4.2 and 4.3).

It is rather less obvious that figures from published papers are also not in the best form for a presentation. Unless they make a key point, figures can take up a lot of space. In some cases, therefore, it may be simpler to present summary results in text form. The figures in the example report in Box 4.2, for example, could, if pushed for time or space, be converted from scattergraphs to the summary text format below:

bigger crickets win more often	$r_s = 0.81 \; n = 24^{***}$
and initiate more fights	$r_s = 0.63 \; n = 24^{*}$
and similarly sized crickets are more likely to escalate	$r_s = -0.71 \; n = 24^{**}$

However, in general, graphs make differences or trends easier to assimilate at a glance and are probably preferable where constraints allow.

Legibility

The ease with which poster and projected text can be read is affected by many things, some of which are specific to the venue and the local conditions. However, there are a few basic points worth bearing in mind:

Font sizes and styles. For slides used in talks, a sizeable font is needed to ensure text can be read comfortably throughout a lecture venue. Our suggestion would be somewhere in the range of 18 pt (for small rooms) to 28 pt (for large theatres), with 24 pt probably being a good all-round working medium. For posters, 14 pt is probably the minimum to go for; where posters are produced in PowerPoint (*see* Box 4.4 *below*), and subsequently blown up to A0 size, the font size on screen is likely to be somewhat smaller.

But font *size* is not the only consideration. Many people have various forms of visual impairment or reading difficulties such as dyslexia. The legibility of text to such people can be greatly affected by the *style* of font chosen. In general, sans serif fonts, like *Arial* or *Verdana*, are much clearer to people with visual impairments or dyslexia than serif fonts such as *Times New Roman*, because of the crisper, cleaner lines of the letters, which give them a sharper boundary with the background.

Contrast and colour combinations. Contrast between text and background is obviously another issue, as is the degree of comfort in reading it. Black text on a white background has high contrast, but it can be rather boring and harsh on the eye in long presentations. Colour combinations such as pale blue or yellow text on a dark blue background, or yellow text on green can be more comfortable, but be careful to avoid combinations that create problems for colour-blind readers (combinations of red and green being an obvious one to avoid – about 8 per cent of males are red/green colour blind). If you want to check the latter, there are various websites and packages that can simulate the appearance of your text to a reader who is colour-blind (e.g. **www.vischeck.com**/).

Getting attention

In both talks and posters, it is necessary to get the attention of people you want to address. You may have a somewhat more captive audience for a talk, but that doesn't stop people's attention wandering, and, if it wanders sufficiently, they may even walk out. There are various things to think about here. One of the first is your title slide or poster header.

The title of a written paper often tends to be literal and rather pedestrian, though people do liven them up with snappy phrases. Thus, the title of a written paper based on the cricket study in Box 4.2 might be something like.

The Effect of Body Size on the Escalation of Aggressive Encounters between Male Field Crickets *Gryllus bimaculatus*

W. G. Grace
Animal Behaviour & Ecology Research Group, School of Biology,
University of Nottingham, University Park, Nottingham NG7 2RD, UK

This is perfectly descriptive of what is to follow, but a lot of text and not very inspiring visually. In the case of a poster, where people are drifting casually past waiting to be hooked by something attractive and interesting-looking, this is unlikely to have much impact. Even in a talk it might suggest a rather dry and cluttered offering is in prospect. Thus it is worth thinking of something catchy as a title and choosing font styles and text layouts that catch the eye. Where fighting in crickets is concerned, something along the lines of the following might do the trick:

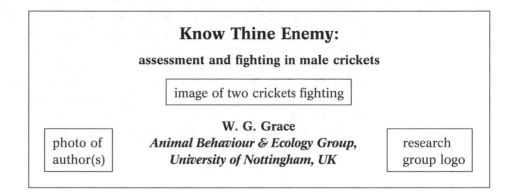

Again, some carefully chosen images can be used to brighten it up.

Gimmicks. In the scramble for attention, there is often the temptation to descend into gimmickry. The advent of PowerPoint (*see* Box 4.4 *below*) has unfortunately brought countless opportunities for this in the slides of spoken presentations, with sound effects, innumerable modes of slide transition, animated cartoons, video windows and many, many more offerings to tempt the over-enthusiastic. While we certainly wouldn't advise eschewing these entirely (some slide transitions, animations and especially the use of video are extremely effective), use them with a great deal of restraint. A gimmicky slide transition with sound effects, for instance, may add a helpful element of surprise the first time, but it will pall to a serious irritation by the third or fourth. In general, the use of gimmicks in slides risks making you look more of a nerd than a slick professional.

To some extent, the case for a bit of carefully judged gimmickry is more arguable in the case of posters, which have to speak out from a background of countless rivals. Most posters are now printed as a single sheet (A0 size) from a file produced by a presentation package such as PowerPoint. However, the advantages of this in terms of ease of mounting and not having to keep track of many separate components can be offset by the greater difficulty of making your poster distinctive. One way in which you can combat this A0 uniformity problem is to have attachments. We have seen convincing recreations of a corner of reed bed (complete with nest and artificial eggs of the study species), clockwork models and miniature sound and video playback systems – all devices that successfully made that poster stand out from the crowd enough to draw us in and begin reading it. But we have also seen efforts (e.g. attached helium balloon banners and clumsy attempts at interactive gadgets) that made the poster look like a tawdry fairground attraction and were decidedly off-putting. Alternative, and quite useful, extras are contact details, abstracts and preprints of the poster on A4 sheets offered in a manner that can be taken away by interested visitors to the poster – usually from a container attached to the poster.

Signposting

In a talk, slides are presented in sequence, so you might imagine there is no problem for the audience in following the logical flow of text and images. This is a dangerous assumption. It is easy to be fooled by your own familiarity with the material into assuming too much on the part of the audience. It is *always* worth having a slide that outlines the structure of your talk and signposts it for people. Thus the famous dictum 'Stand up, say what you are going to say, say it, then sit down' is excellent advice. By signposting, of course, we don't mean listing the conventional sections of a written paper (Introduction, Methods, Results, etc.) Even though your talk may *de facto* follow these sections, you should use the opportunity of a talk structure slide to give a more specific indication of what the talk will be about. For instance, to pursue our cricket example, a structure slide here might go something like:

Fighting and assessment in crickets

1. why assess opponents in aggressive contests?
2. aggression in crickets
3. experimental design
4. the evidence for assessment
5. modelling assessment strategies
6. testing the model's predictions
7. conclusions and the wider view

Again, this can be embellished with a judiciously chosen image in the corner to give it some visual interest. The structure slide can be reprised at various points in the talk to update the audience on where things have got to and where they're about to go.

A second piece of advice in terms of structuring a spoken presentation is to move on to a new slide or point in an animated list (*see* Box 4.4 *below*) only when you are ready to deal with the new material. To judge by the many speakers who move on to the next slide or point before finishing the current one, there is a strong temptation to hurry on. Resist the temptation! The audience's attention will be taken by the new material, your current point will be lost, and confusion is likely to start setting in.

Signposting in a poster presentation. In a poster, text and images tend to be presented in blocks spread over the area of the presentation. Under these circumstances it can sometimes be difficult to determine the intended route through it all. Two obvious ways of overcoming this are to use arrows between successive pieces of text, or to number each piece of text in sequence. Associated images, tables and figures can then be numbered as Plate n, Table n or Figure n accordingly. Numbering blocks of text is probably neater; arrows running through posters can sometimes make them look fussy and cluttered, and you can always use a different coloured font for your numbers to make them stand out.

Summaries

In many ways almost the most important part of your presentation is the final impression you leave with your audience/readers. You need to finish with a bit of a bang in the form of some clear, crisp take-home messages that will stay with those hearing or reading them. Thus, it is vital that you don't just peter out at the end of a rambling discussion and leave the audience/readership to decide for itself what was important about the work. Present a clear *summary* slide or poster panel! The summary should be succinct and memorable, and, in a talk, we strongly advise making it the *last* slide (rather than the common practice of ending with the equivalent of the Acknowledgements section of a written paper). Not only is it then the last slide the audience sees, but it can be displayed for the longest period as it often remains on display during the question session. A summary slide/panel for the study of fighting in crickets might go something like:

Summary

- Large male *G. bimaculatus* are more likely to win fights
- Fighting is more severe when opponents are closely matched in size
- Both prior assessment and individual confidence influence the decision to fight
- Fighting in the absence of females or reproductive resources may help reduce competition when they are available

---

Questions

Questions from your audience/readers are an integral part of both spoken papers and posters and deserve as much consideration as the talk or poster itself. No study is ever the final word, and other people will often have refreshingly new views on how it might be improved or developed. It is often possible to imagine the sort of questions you could be asked, because you will be aware of the study's shortcomings and what remains to be done yourself, and it is well worth thinking through these ahead of your presentation. But by far the best way to prepare for what might come is to do some dummy runs of your talk or poster to groups of people, such as classmates, members of your research group, departmental seminar groups or whatever's handy. This almost always throws up issues you won't have thought of, and is particularly good at highlighting where you're failing to make things clear. When on the spot receiving questions at your presentation proper, listen carefully to what is actually being asked; under stress there can be a tendency to hear the question you are expecting rather than what is really being asked. One way to avoid this trap is to begin a reply by briefly paraphrasing the question back to the asker; not only does this help to establish you are answering the right question, but it also gives a little extra time to marshal your thoughts for the answer.

4.5.3 Using Microsoft PowerPoint to prepare slides and posters

Just as with preparing text and organising and analysing data, there are commercial software packages designed for the world of spoken and poster presentations at meetings. For the same reasons as Excel (flexibility and availability), Microsoft's PowerPoint is the almost universal package of choice here. However, just like Excel, PowerPoint offers a combination of rich opportunities and decided pitfalls. As with Excel, we don't have the space here to give a full-blown tutorial on using PowerPoint, but what follows below and in Box 4.4 is a brief introduction to the package's pluses and minuses.

The most obvious production advantage of PowerPoint is that slides for spoken presentations are produced electronically and not photographically (i.e. as transparencies or diapositives), a process associated with time delays, inflexible results and, often, high cost. Unfortunately, the same advantage doesn't apply to posters. Although PowerPoint allows posters to be created electronically (*see* Box 4.3), the actual poster still has to be produced through a time-consuming and costly printing process and there is little scope for tinkering without reprinting the entire thing.

One of the most irritating, and potentially disastrous, problems with PowerPoint is that, despite being almost ubiquitous, different versions vary considerably and perform differently on different computer systems, data projectors and printers. As a result, there is no guarantee that what you saw during the creation process will appear during your talk or in your final poster, and, with depressing regularity, your laptop may not work with the particular data projector in the auditorium. Even when all the hardware works together, colours, fonts, layouts, animation effects and audio/visual embellishments are common casualties of different versions of the software. *It is therefore essential to check your presentation thoroughly on the system that will be used at the time well ahead of the giving it for real*. It is also worth checking the compatibility of transfer media ahead of a talk; many times we have seen the miserable consequences of discovering five minutes before being introduced that the host machine won't read the CD, USB stick or whatever on which the speaker has saved their presentation, or won't connect with the speaker's home network on which it resides. Problems in all respects are particularly likely if you are switching between PC and Macintosh systems, so always check doubly thoroughly here. Many Mac users convert their PowerPoint presentation into a pdf file[†] and use the *full screen view* option (on the Window menu of the Adobe Acrobat programme) to display the image. Often the pdf file is considerably larger than the original. A further disadvantage is that animated transitions within a slide (see *'Build' effects within slides*, p. 196) are not preserved when the pdf file is created. This can be overcome by making several versions of the same slide in PowerPoint and successively adding details before creating the pdf.

With such technical considerations in the background, Box 4.4 introduces some basic facilities offered by PowerPoint for creating slides and posters, but personal experience with using it will lead you to lots of others and your own personal preferences for style and approach.

[†] This requires that Adobe PDF appears as a printer option when the Print option is chosen from the File menu of PowerPoint.

Box 4.4	A basic guide to PowerPoint

Microsoft's PowerPoint is now almost the default vehicle for preparing and giving spoken and poster presentations. What follows is an extremely basic summary of some of its facilities in this respect. As with most packages, there are usually several ways of achieving the same result (e.g. the keyboard short-cut Control Z undoes an action, as does selecting 'Undo' from the drop-down Edit menu). Most users quickly become familiar with one way of doing things and often remain in blissful ignorance of the others.

Preparing slides for a talk

Slide styles

When you first boot up PowerPoint the screen offers a default slide format that requests a title and subtitle. You may wish to use this, or you may want a different format. If the latter, click on the 'New' (empty page) icon on the far top left of the screen and a range of alternative slide layouts will appear on the right of the screen (of which the boot-up default layout is the first). Single click on the design you want and it will appear centre screen ready for completion. Where styles have ready-formatted text boxes, such as the first four in the 'Test Layouts' panel, text can be added simply by clicking in the box and typing in. When using the blank open slide in the first 'Content Layouts' format, text must be added by clicking on the 'Text Box' button (the lined panel with the 'A' in the top left corner) on the bottom tool bar, and immediately clicking on the slide to create a panel within which to type text. Of course, the open slide format can be used for many other things besides text, including graphs, photographs, tables, video windows, free-style drawing and so on. Other formats offer different combinations of image, table and text as indicated.

Backgrounds

The form and colour of the background of a slide can be changed by clicking on 'Format' at the top of the screen, followed by 'Background' and selecting the various colour and fill pattern options on offer. Remember to bear in mind the legibility of any text you may want to overlay on the background. In general, the simpler and plainer the background the better. Complicated and fussy images, like photographs of scenery or organisms, usually make very poor, confusing backgrounds.

Font size and style

The font size can be changed by selecting the appropriate number in the 'Font' box in the top tool bar (the white box with a number in it next to the font style [e.g. Arial, Times New Roman] option). Follow the guidelines for font sizes and styles in the main text. The colour of your text can be changed by highlighting it and clicking on the A with a coloured (default black) bar underneath it on the tool bar and choosing a colour from the available options; alternatively click on the symbol and choose a colour prior to typing your text. Be aware of the strengths and pitfalls of different combinations of font and background colours (*see text*).

Transitions between slides

PowerPoint offers many different transition effects when moving from one slide to the next: fade or dissolve in, wipe effects, cover over and many more. To choose one, click on the 'Slide sorter view' button (the one with four small squares in it) at bottom left of the screen. This displays your presentation as rows of slides. Now click on the slide(s) (hold down 'Shift' if you wish to select more than one slide simultaneously) for which you wish the transition effect to apply, then click on 'Transition' in the top tool bar. This produces a menu of options to the right of the screen, on which you can click to make your selection. PowerPoint will briefly demonstrate the effect as you click it.

'Build' effects within slides

As well as customising transitions between slides, you can customise the manner in which text appears in some slide formats. The usual format for this is the numbered or bullet-pointed text slide (the third in the panel of 'Text Layouts' formats above). To choose a style here, click on the 'Normal view' button to the farthest left at the bottom of the screen and move to the slide you wish to customise. Now click on 'Design' in the top tool bar, then on 'Animation schemes' in the options panel that pops up to the right, and finally on your chosen effect. Again PowerPoint will quickly demonstrate the effect.

Images and movie and sound files

Images, such as photographs or graphs or tables from statistics packages or scanned from publications, can easily be added to slides in PowerPoint, either from the Windows clipboard, if they've been copied there from another Windows source (just use the 'Paste' option to copy them into the active slide in 'Normal view'), or using the 'Insert' option, if the images are in files saved elsewhere. To use the latter, click on 'Insert' and, if you want, say, a photograph or figure that is saved as a 'jpg' or 'gif' file, click 'Picture' and 'From File' then browse for the file concerned. Double-click on the file name and the image will appear in the slide. You will now need to size and position it as required using the mouse. The 'Picture' option also allows other possibilities, such as inserting 'Clip Art' images and 'Autoshapes' (the standard drawing shapes available from the lower Windows tool bar).

Movie files (such as 'avi' or 'mpeg' files) can be inserted in the same way. From 'Insert' click on 'Movies and Sounds', then 'Movie from File' and browse for the required file. Double-click on it to make it appear in the slide. At this point PowerPoint gives you the option of having the movie play automatically or not until you click on it. Choose which you wish, and size and position the movie window as required. You can now customise the animation in various ways. To make the movie play as soon as the slide appears, right-click on the movie window and choose 'Custom Animation'. In the right-hand column that now appears, click on the downward pointing arrowhead on the blue background immediately next to the title of the movie file and check the 'Start With Previous' option. Click on the arrow again, select 'Timing' and make sure there is a zero second delay (or type in a delay of your choice); then click 'OK'. If you want the movie clip to repeat play until you have finished with the slide, right-click on the movie window again and click on 'Edit Movie Object'. In the resulting dialogue box, check the 'Loop until stopped' option and click 'OK'.

For sound (e.g. 'wav') files, the process is the same as for movies up to the point that a small loudspeaker icon appears on the slide to show that a sound has been inserted. One very useful feature is that by highlighting this icon and using 'Custom Animation', you can make the sound play when you press the Enter key, avoiding the need to position the mouse arrow over the icon. You can also arrange the slide in an animation sequence triggered by the appearance of a previous item such as a picture of the animal making the sound. An important consideration for both sounds and movies is the size of the file. Even short clips usually exceed 1 MB and take what seems minutes to load. Remember to keep clips as short as possible and check how long they take to load – preferably with the setup that will be used when delivering the talk.

Checking slides and running the presentation

At any point during the creation of your presentation, you can check what slides will look like when actually running by clicking on the 'Slide Show' button, which is the one with the goblet-like icon representing a slide screen on the bottom tool bar. This is also the button to click when you give the presentation itself. Left-clicking the mouse, or pressing the 'Page Down' or carriage return keys on the keyboard, will move the slide show forwards. Right-clicking the mouse and clicking 'Previous', or pressing the 'Page Up' key, will move the slide show back.

Preparing a poster

A poster in PowerPoint is usually prepared as a single slide, which is then printed at whatever size is required, normally A0, but A3 and other sizes are occasionally used. Because you are working within a single slide, it is usually necessary to work at something like 150 per cent or 200 per cent normal size, which can be set using the percentage 'Zoom' scale in the top tool bar. Adding text is then a matter of clicking on the 'Text Box' button on the bottom tool bar and immediately clicking on the slide to create a panel within which to type text (as for open field slides above). Because you are working at magnification, there is a certain amount of judgement to be made about font sizes. First you need to decide how much text there is going to be in total, and how this will be broken up and mixed with figures, tables and images across the poster. Then choose a font size that will accommodate this, but be clearly legible at a comfortable distance once the poster is printed up. You can usually judge this from its appearance in 'Slide Show' (*see above*) format. Graphs, tables, photographs, etc. can be inserted as for talk slides via the clipboard or 'Insert' options. Text boxes and images can then be moved around with the mouse until the desired layout has been achieved. Don't forget a good clear banner or box title with all the required information in large, clear font (see suggestions for title slides in the text).

4.6   Plagiarism

Before leaving our discussion of reporting work, we have to mention one more issue that has recently been gaining attention in the general education community, and that is *plagiarism*. Plagiarism refers to any attempt to pass off somebody else's work as your own, and encompasses a wide range of possibilities from downloading pre-prepared essays from the Internet/World Wide Web, or copying-and-pasting tracts of text or other material from other people's documents into your own without due acknowledgement of the source, to the slightly greyer area of insufficient referencing or use of quotation marks where odd phrases, sentences or ideas have been gleaned from the literature. Science depends on honesty and an ability to attribute ideas and results properly in reporting studies, so quite apart from any dishonest advantage that may be sought in educational assessments, wanton plagiarism undermines the scientific process itself. It is therefore a serious matter. Of course, the possibility of plagiarism has always been with us, but, in recent years, the Internet/World Wide Web has greatly exacerbated the problem by making available a vast amount of material that is downloadable at the push of a button and (at least until very recently) difficult for an external party to trace. Many educational establishments now use sophisticated anti-plagiarism software, which can scan a submitted piece of work and quickly match its contents to other electronic sources, so the arms race against cheats has moved on significantly. However, as with any policing process, it is possible to fall foul of the system inadvertently through carelessness or ignorance. Our advice, therefore, is to take careful note of the reporting conventions in professional journals, books and conferences, and, especially if you're a student, the guidelines given to you by your university, college or school.

4.7   Summary

1. Confirmatory analyses are usually presented in summarised form (e.g. summary statistics, scattergrams) as tables *or* figures *or* in the text of a report. In all cases, sample sizes (or degrees of freedom), test statistics and probability levels should be quoted. In the case of tables and figures, these can be included within the table or figure itself or within a full, explanatory legend.

2. Results should almost never be presented as raw numerical data because these are difficult for the reader to assimilate. In the exceptional circumstances where the presentation of raw data is helpful, presentation should usually be selective to the points being made and is best incorporated as an Appendix.

3. The axes of figures should be labelled in a way that conveys their meaning clearly and succinctly. Where analyses in different figures are to be compared directly, the axes of the figures should use the same scaling.

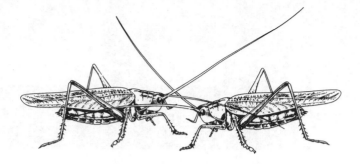

4. The legends to tables and figures should provide a complete, self-contained explanation of what they show without the reader's having to search elsewhere for relevant information.

5. Reports of investigations should be structured into clearly defined sections: Introduction, Methods, Results, Discussion, References. Each section has a specific purpose and deals with particular kinds of information. The distinction between them should be strictly maintained. When reports are long or involved it can be helpful to add an Abstract and/or Conclusions section to highlight the main points and take-home messages.

6. At a time when increasing importance is being attached to the public understanding of science, writing for a general interest readership, as opposed to the more usual professional scientific one, is a useful skill. Success often depends on avoiding technical jargon, or at least explaining it in everyday language, and using familiar analogies or imagery to get attention and facilitate understanding.

7. Information can also be presented in the form of talks or poster papers at meetings. This can be a very important means of disseminating your work and getting yourself known, and, as usual, there are various 'dos' and 'don'ts' that make for good practice in preparing presentations.

8. Be sure to reference sources of information properly in any presentation of your work. Plagiarism (an attempt to pass off somebody else's work as your own) is a serious offence in education and scientific research.

Reference

Sokal, R. R. & Rohlf, F. J. (1995) *Biometry*, 3rd edition. Freeman, San Francisco.

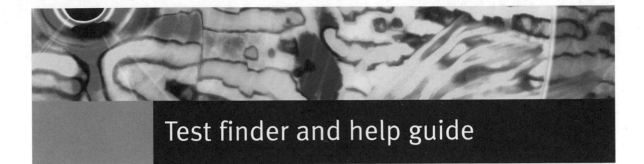

Test finder

A   Looking for a difference or a trend? (if unsure, go to *Help* 1)
    * Trend   – go to B
    * Difference   – go to C

B   Have you fixed the X-values experimentally (looking to see whether changes in X cause changes in Y) or do you require the equation of the line itself (e.g. for predicting new values)? (if unsure, go to *Help* 2)
    * No   – correlation
    * Yes   – linear regression

C   Does each data value belong to treatments within one level of grouping, or does it belong to more than one level of grouping? (if unsure, go to *Help* 3 and *Help* 6)
    * One level of grouping (one-way designs)   – go to D
    * More than one level of grouping (*n*-way designs)   – go to G

D   Are data for each treatment replicated? (if unsure, go to *Help* 4)
    * No   – go to E
    * Yes   – go to F

E   Data are in the form of counts
    * No   – (not analysable)
    * Yes   – $1 \times n$ chi-squared

F   Data in one treatment are independent of those in other treatments (if unsure, go to *Help* 5)
    * No   > only two treatments   – matched-pairs tests, or use one-way repeated-measures ANOVA

           > two or more treatments   – one-way repeated-measures ANOVA
    * Yes   > only two treatments   – *t*-test or Mann–Whitney test or use one-way ANOVA

           > two or more treatments   – one-way ANOVA

G   Are data for treatments within each level of grouping replicated? (if unsure, go to *Help* 6)
    * No, data are counts   – $n \times n$ chi-squared
    * Yes   – *n*-way ANOVA

Help 1

Difference or trend?

■ *Difference* predictions are concerned with some kind of difference between two or more groups of data. The groups could be based on any characteristic that can be used to make a clear-cut distinction, e.g. sex, drug treatment, habitat. Thus a difference might be predicted between the growth rates of men and women, or between the development of disease in rats given drug A versus those given drug B versus those given a placebo.

■ *Trend* predictions are concerned not with differences between mutually exclusive groupings but with the relationship between two more-or-less continuously distributed measures, e.g. the relationship between the size of a shark and the size of prey it takes, or the relationship between the amount of rainfall in a growing season and the number of apples produced by an apple tree.

Help 2

What sort of trend?

The basic choice here is between fitting a line (regression) or not (correlation).

Correlation is used whenever we merely want to know whether there is an association between X and Y – there is no real dependent or independent variable, and when plotted as a scattergraph you could equally well plot Y against X as opposed to X against Y.

Regression was developed for situations where the experimenter manipulates the values of the *independent* variable (X) and measures the impact of these manipulations on another *dependent* variable (Y). You are therefore looking for a causal relationship between X and Y – changes in X cause changes in Y. However, because knowing the slope of the relationship between X and Y is useful in many other contexts, the use of regression has expanded to incorporate many cases where the X values are merely measured rather than manipulated. Technically this is wrong, but this usage is so firmly embedded in biological practice that the majority of scientific investigations do it. An example is when you want to use the relationship to predict the value of Y for a particular X value. Comparisons of slopes are also extremely informative, and used extensively in biological research.

Help 3

Levels of grouping

Many difference predictions are concerned with differences at just *one* level of grouping, e.g. differences in faecal egg counts following treatment of mice with one of four different anthelminthic drugs. Here drug treatment is the only level of grouping in which we are interested. However, if we wished, say, to distinguish between the effects of different drugs on male and female mice, we should be dealing with *two* levels of grouping: drug treatment and sex.

Help 4

Replication

Replication simply means that each treatment within a level of grouping has more than one data value in it. The table below shows replicated data as columns of values within each treatment in a one-way (i.e. one level of grouping) design:

	Group		
	Pesticide A	Pesticide B	Control
% mortality of pest	10	30	0
	5	27	1
	3	50	1
	0	6	0
	1	3	2
	20	3	5

Help 5

Independence

Unless specifically allowed for in the analysis, all statistical analyses assume that each data value is independent of all others. Each value in one group is *non-independent* of one from each of the other groups if they have something in common (e.g. it is measured on the same individual, or derives from animals from the same cage – i.e. a given individual is exposed to each treatment in turn, or the same cage provides animals for each treatment). The source of non-independence thus needs to be taken into account in any analysis, and so requires a so-called repeated-measures design.

Help 6

Rows and columns

If data have been collected at two levels of grouping, then each data value can be thought of as belonging to both a row and a column (i.e. to one row/column cell) in a table, where rows refer to one level of grouping (say sex – *see* Help 4) and columns to the other (drug treatment – *see* Help 4). If there are several values per row/column cell, as below for the number of individuals dying during a period of observation:

		Treatment	
		Experimental	Control
Sex	Male	3, 4, 8, 12	23, 24, 12, 32
	Female	1, 0, 2, 9	32, 45, 31, 21

then a two-way analysis of variance is appropriate. If there is just a single *count* in each cell, as in the number of male and female fish responding to an experimental or control odour stimulus:

		Treatment	
		Experimental	Control
Sex	Male	27	91
	Female	12	129

then an $n \times n$ chi-squared test is appropriate.

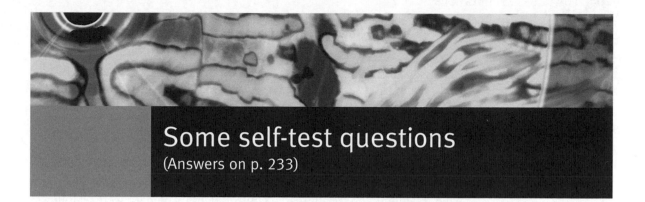

1. An experimenter recorded the following body lengths of freshwater shrimps (*Gammarus pulex*) in three different lakes.

Body size (mm)		
Lake 1	Lake 2	Lake 3
9.9	10.5	9.6
8.7	12.1	9.0
9.6	11.2	8.7
10.7	9.7	13.2
8.9	8.7	11.9
8.2	11.1	14.0
7.7	10.7	12.9
8.1	11.8	10.8

Faunal diversity in the lakes was known ($1 < 2 < 3$) and the experimenter expected shrimps from more diverse lakes to be smaller because of increased interspecific competition. To test this idea he compared body lengths in each pair of lakes (1 versus 2, 2 versus 3 and 1 versus 3) using Mann–Whitney *U*-tests. Is this an appropriate analysis? If not, what would you do instead?

2. How would you decide between correlation and regression analysis when testing a trend prediction?

3. The following is part of the Discussion section of a report into the effects of temperature and weather on the reproductive rate of aphids on bean plants.

   While the results show a significant increase in the number of aphids produced as temperature rises, there is a possible confounding effect of the age of the host plant and the rate of flow of nutrients. Indeed, there was a stronger significant positive correlation between nutrient flow rate and the number of aphids produced ($r_s = 0.84$, $n = 20$, $p < 0.01$) than between temperature and production (*see* Results).

   Do you have any criticisms of the piece?

4. What does the following tell you about the analysis from which it derives?

$$H = 14.1, \qquad \text{d.f.} = 3, \qquad p < 0.01$$

5. An agricultural researcher discovered a significant positive correlation ($r_S = 0.79$, $n = 112$, $p < 0.01$) between daily food intake and the rate of increase in body weight of pigs. What can the researcher conclude from the correlation?

6. A plant physiologist measured the length of the third internode of some experimental plants that had received one of three different hormone treatments. The physiologist calculated the average third internode length for each treatment and for untreated control plants. The data were as follows:

	Control	Hormone 1	Hormone 2	Hormone 3
		Treatment		
Average internode length (mm)	32.3	41.6	38.4	50.2

To see whether there was any significant effect of hormone treatment, the physiologist performed a 1×4 chi-squared test with an expected value of 40.6 in each case and three degrees of freedom. Was this an appropriate test?

7. What do you understand by the terms:

(a) test statistic,
(b) ceiling effect,
(c) statistical significance?

8. The figure shows a significant positive correlation, obtained in the field, between body size in female thargs and the percentage of females in each size class that were pregnant. From this, the observer concluded that male thargs preferred to mate with larger females. Is such a conclusion justified? Give reasons for your answer.

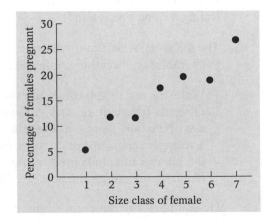

9. Why are significance tests necessary?

10. The following were the results of an experiment to look at the effect of adding an enzyme to its substrate and measuring the rate at which the substrate was split. In the control Treatment A, no enzyme was added; in Treatment B, 10 mg of enzyme was added; in Treatment C, 10 mg was added but the reaction was cooled; and in Treatment D, 10 mg was added but the reaction was warmed slightly.

Treatment A	Treatment B	Treatment C	Treatment D
0	20	52	71
1	21	69	92
2	35	100	55
1	15	32	78
0	20		105
0	24		82
			92

What predictions would you make about the outcome of the experiment and how would you analyse the data to test them?

11. A farmer called in an agricultural consultant to help him decide on the best housing conditions (those resulting in the fastest growth) for his pigs. Three types of housing were available (sty + open paddock, crating, and indoor pen). The farmer also kept four different breeds of pig and wanted to know how housing affected the growth rate of each. What analysis might the consultant perform to help the farmer reach a decision?

12. Figures (a) and (b) were used by a commercial forestry company to argue that the effect of felling on the number of bird species living in managed stands (assessed by a single standardised count in each case) was similar in both deciduous and coniferous forest. Would you agree with the company's assessment on this basis?

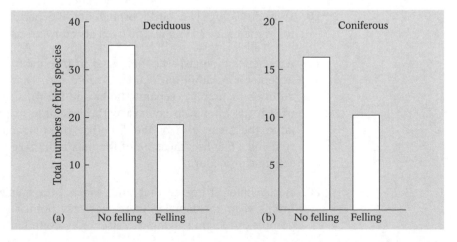

13. What would a *negative* value of the test statistic r signify to you?

14. Derive some hypotheses and predictions from the following observational notes:

> Sampled some freshwater invertebrates from three different streams using a hand-net. There were more individuals of each species at some sites than others, both within streams and between them. Also some sites had a more or less even distribution of individuals across all species whereas others had a highly biased distribution with some species dominating the community. Some species occurred in all three streams but they tended to be smaller in some streams than others. A number of predatory dragonfly nymphs were recorded but there was never more than one species in any one sample, even when more than one existed in a stream. Water quality analyses showed that one stream was badly polluted with effluent from a local factory. This stream and one of the others flowed into the third stream, forming a confluence. It was noticed that stones and rocks on the substrate had fewer organisms on or under them in regions of faster flow rate.

15. Is chi-squared used for testing differences or trends?

16. A student in a hall of residence suffered from bed bugs. During the course of a week he was bitten 12 times on his legs, 3 times on his torso, 6 times on his arms and once on his head. Could these data be analysed for site preferences by the bugs? If so, how?

17. An experimenter had counted the number of times kittens showed elements of play behaviour when they were in the presence of their mother or their father and with or without a same-sex sibling. The experimenter had collected ten counts for each condition: (a) mother/sibling, (b) mother/no sibling, (c) father/sibling, (d) father/no sibling, and was trying to decide between a 2×2 chi-squared analysis and a 2×2 two-way analysis of variance. What would you suggest and why?

18. Why do biologists regard a probability of 5 per cent or less as the criterion for significance? Why not be even stricter and use 1 per cent?

19. A fisheries biologist was interested in the maximum size of prey that was acceptable to adult barracuda. To find out what it was, he introduced six adult barracuda into separate tanks and fed them successively larger species of fish (all known to co-exist with barracuda in the wild). He then calculated the mean size of the fish that the barracuda last accepted before refusing a fish as a measure of the maximum size they would take. Is this a sensible procedure?

20. A psychologist argued that since males of a species of monkey had larger brains than the females there was less point in trying to teach females complex problem-solving exercises. Any comments?

21. An ecologist studying populations of voles in different woods suspected from a glance at the data that males from some woods had larger adrenal glands than those from others. Unfortunately, the age of the animals also appeared to differ between the woods. How might the ecologist test for a difference in adrenal glands between woods while controlling for the potential confounding effect of age?

22. What do you understand by an 'order effect'?

23. A parasitologist wished to test for the effect of increasing the amount of food supplement to nestling birds on subsequent parasite burdens as adults. The parasitologist intended to carry out a linear regression analysis with amount of food supplement on the x-axis and adult parasite burden on the y-axis. However, when the distribution of parasite burden data was checked for normality, the following was found:

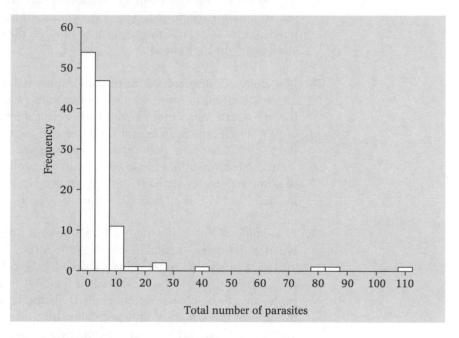

What implications does this have for the regression analysis?

24. What is meant by pseudoreplication?

25. When a botanist compared the frequency of different leaf sizes on a tree with a normal distribution, the significance test comparing the two distributions yielded a test statistic value with an associated probability of 0.0341. Do the botanist's data conform to a normal distribution or not?

26. A behavioural ecologist was interested in the effects of other competing males in the environment on sperm production by focal courting males. Her prediction was that courting males should produce more sperm to transfer to a female if there are more males around to compete for her. She decided

to test her idea by exposing 20 focal males to each of three treatments: (a) no other males present, (b) one other male present, and (c) four other males present. She was careful to randomise the order in which the three treatments were presented to her different subjects and she measured the width of the thorax of each subject male and stimulus female to gauge their body size. The number of sperm produced was estimated as the weight of the spermatophore (package of sperm) transferred to the female. How might she go about analysing the data from the experiment to test her prediction?

27. A psychologist interested in sexual attitudes in men and women from different social environments carried out a questionnaire survey of some 150 subjects. The questionnaire consisted of 40 questions covering a range of aspects of sexual behaviour from behavioural characteristics preferred in potential partners to attitudes towards promiscuity and homosexuality. It also asked for various items of background information, such as financial status, family background, age, religion and ethnic group. How might the psychologist set about preparing this kind of data for analysis, and what analysis might they choose?

28. In a study of the effects of various environmental variables on resistance to an experimental infection in male house mice (*Mus domesticus*), a student had measured four variables with respect to subject males: food availability during development, maternal body weight, local population density and litter size. These were then related to a measure of resistance to an experimental infection with a blood protozoan (clearance rate) by means of four separate regression analyses, with clearance rate as the dependent variable in each case. Is this a reasonable approach to the analysis?

29. In a study of the behaviour of foraging bees, individual bees were presented with flowers treated in one of three different ways: (a) not previously visited by any bees, (b) previously visited by the subject forager herself, and (c) previously visited by a different individual forager. Several aspects of each subject's behaviour were measured, including the speed of approach to the flower, the length of time spent hovering in front of the flower before visiting it, whether the flower was touched by the bee, how many times the bee touched the flower with its antennae, the duration of antenna touches, whether the flower was visited at all, the length of any visit, the length of time spent hovering by the flower after a visit, and the speed of departure from the flower. How might these data be analysed to reveal how the bees responded to the different flower treatments?

30. A physiologist was interested in the effect of mobile phones on nerve cell (neurone) function. He therefore measured impulse transmission rates in replicated laboratory preparations of neurones under three different strengths of electromagnetic field, plus a control of zero magnetic field. Each preparation was only used once, and each treatment was replicated 30 times. What kind of analysis is appropriate?

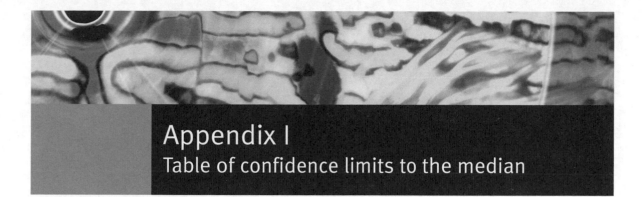

Appendix I
Table of confidence limits to the median

Table of non-parametric confidence limits to the median

Sample size (n)	r (for p approx. 95%)
2	–
3	–
4	–
5	–
6	1
7	1
8	1
9	2
10	2
11	2
12	3
13	3
14	3
15	4
16	4
17	5
18	5
19	5
20	6
21	6
22	6
23	7
24	7
25	8
26	8
27	8
28	9
29	9
30	10

r denotes the number of values in from the extremes of the data set that identifies the 95 per cent confidence limits (*see text*). Modified after Colquhoun (1971) *Lectures on biostatistics*, Clarendon Press, Oxford.

Appendix II
How to calculate some simple significance tests

Examples of tests for a difference between two or more groups

Box A1	The basic calculations for parametric test statistics

If you calculate the test statistic of many parametric statistical tests by hand, the calculations often involve a standard set of core operations on the data. Rather than repeating them in all the relevant boxes, we have placed them here for reference. They all involve calculating a quantity known as the *sum of squares*, which is actually the same operation performed when finding the standard deviation or standard error (*see* Box 2.1).

1. Follow steps 1–3 of Box 2.1. The resulting quantity is the *sum of squares of x*, conventionally denoted by S_{xx}.

Sometimes you will have a *y*-variable as well as an *x*-variable (e.g. in regression, *see* Box 3.11). If so:

2. Calculate the sum of all the *y*-values in the data set (Σy).

3. Square all the *y*-values and sum them, giving (Σy^2).

4. Multiply the *x*- and *y*-values of each *x*–*y* pair together, and add them together, giving (Σxy).

5. Calculate $\Sigma y^2 - [(\Sigma y)^2/n]$, giving the *sum of squares of y*, S_{yy}.

6. Calculate $\Sigma xy - [(\Sigma x)(\Sigma y)/n]$, giving the *sum of the cross-products*, S_{xy}.

Box A2a	(i) Mean values: how to do a general parametric test for two groups (two-tailed *t*-test)

1. Frame the prediction. In this case it is the general prediction that the two groups (A and B) will have different mean values (A ≠ B). Thus the null hypothesis is that the two group means will not differ.

2. Count the number of data values in the first group; this number is referred to as n_1. If $n_1 = 1$, then there is definitely something wrong! (If this number represents a count of the number of items in one of two categories – that form the two

groups – then you should be doing a χ^2 test. If it is a rank or a constant interval measurement, then you need to collect some more data!) Count the number of data values in the second group; this number is referred to as n_2 (it should be greater than 1, as before).

3. Calculate the variances of each group separately (*see* Box 2.1 for how to do this), producing s_1^2 and s_2^2.

4. Calculate the value of $P = (n_1 + n_2)/(n_1 n_2)$.

5. Calculate the value of $Q = [s_1^2(n_1 - 1) + s_2^2(n_2 - 1)]/(n_1 + n_2 - 2)$.

6. Calculate the value of $R = \sqrt{(PQ)}$. This is the standard error of the difference between the two groups.

7. Calculate the mean values of each group separately, and take the difference $S = (\mu_1 - \mu_2)$. Switch the mean values round so as to make the difference positive, since in a general test we are not interested in the direction of the difference, but merely in whether it differs from zero.

8. Calculate the value of the test statistic, $t = S/R$.

9. In order to calculate t, we needed to know the difference in mean values, and the s.e. of this difference, i.e. two prior parameters were required. The degrees of freedom of t are therefore $n_1 + n_2 - 2$.

10. Look up the two-tailed value of t in Table D of Appendix III for the critical value for your degrees of freedom. If your value is greater than the relevant value in the table, then the difference you found is significant.

Box A2a   (ii) Mean values: how to do a specific parametric test for two groups (one-tailed t-test)

1. Frame the prediction, i.e. decide which of the two groups (A and B) is predicted to have the greater mean value. There needs to be some a priori reason (theory, or previous published or gathered data) for this prediction. Suppose that on the basis of your knowledge, you predict that A > B. The null hypothesis is that the mean values for the two groups do not differ in the predicted direction.

2. Calculate the value of t as steps 2–9 above, but make sure that the difference in mean values is done in the way that is *predicted* to generate a positive difference. Here you are predicting that A > B, and hence (A – B) should generate a positive value of t. If it does not actually generate a positive difference, then you know automatically that the result is not significant. Note that if the result is an unusually large but negative value of t (that would have been significant, had you predicted the opposite pattern of mean values), you are not allowed to conclude *anything* other than that your prediction is not supported by the data. This is the cost of a specific (one-tailed) prediction paid in exchange for the benefit of the more powerful test.

3. Look up the critical value of a one-tailed t-test in Table D of Appendix III using the degrees of freedom you have. If your t-value is greater than the critical one, then you conclude that the result is significant: the evidence suggests that the group mean predicted to be greater really is so, and you reject the null hypothesis.

Box A2b (i) Mean ranks: how to do a general non-parametric test for two groups (Mann–Whitney U-test)

1. Count the number of data values in the group with the fewer values (if there is one); this number is referred to as n_1.

2. Count the number of data values in the other group; this is n_2.

3. Rank all the values *in both groups combined*. The smallest value takes the lowest rank of 1, the next smallest value a rank of 2 and so on. If two or more values are the same they are called *tied values* and each takes the average of the ranks they would otherwise have occupied. Thus, suppose we have allocated ranks 1, 2 and 3 and then come to three identical data values. If these had all been different they would have become ranks 4, 5 and 6. Because they are tied, however, they each take the same average rank of 5 $((4 + 5 + 6)/3)$, though – and this is important – the next highest value still becomes rank 7 just as if the three tied values had been ranked separately. If there had been only two tied values they would each have taken the rank 4.5 $((4 + 5)/2)$ and the next rank up would have been 6. We should thus end up with rank values ranging from 1 to N, where $N = n_1 + n_2$.

4. Add up the rank values within each group, giving the totals R_1 and R_2 respectively.

5. Calculate $U_1 = n_1 \times n_2 + [(n_1(n_1 + 1))/2] - R_1$.

6. Calculate $U_2 = n_1 \times n_2 - U_1$.
   If U_2 is smaller than U_1 then it is taken as the test statistic U. If not, then U_1 is taken as U.

7. We can now check our value of U against the threshold values in U tables (a sample is given in Appendix III, Table B). If our value is *less* than the threshold value for a probability of 0.05, we can reject the null hypothesis that there is no difference between the groups. Note that, in this test, we use the two sample sizes, n_1 and n_2, rather than degrees of freedom to determine our threshold value.

If one of the groups has more than 20 data values in it, U cannot be checked against the tables directly. Instead, we must use it to calculate another test statistic, z, and then look this up (some sample threshold values are given in Appendix III, Table C). The calculation is simple:

$$z = \frac{U - (n_1 \times n_2)/2}{[(n_1)(n_2)(n_1 + n_2 + 1)]/12}$$

Box A2b (ii) Mean ranks: how to do a specific non-parametric test for two groups

There is no non-parametric test specially designed to test for a specific difference between two groups. Use the specific non-parametric analysis of variance (*see* Box 3.4a), because this can cope with any number of groups.

Box A3a   (i)   Mean values: how to do a general parametric one-way analysis of variance (ANOVA)

1. Frame the prediction. In this case, the general prediction being made is to ask whether there are any differences among the mean values of the groups. Therefore the null hypothesis actually tested is that there are no differences among the mean values of the groups.

2. The test considers i groups of data, each of which contains n_i data values. The total number of data values in all groups together $= N = \Sigma n_i$.

3. Calculate the mean values of each of the groups, μ_i.
   Work out T, the total sum of squares of all the data (follow Box 2.1, items 1–3).

4. Work out S_i, the sum of squares for the data of each group separately (again, follow Box 2.1, 1–3).

5. Calculate the error sum of squares, $SS_{error} = \Sigma S_i$.

6. Calculate the among-groups sum of squares, $SS_{among} = T - SS_{error}$. This should give the same result as calculating the sum of squares using the mean values μ_i rather than the raw data.

7. The d.f.$_{total}$, the total degrees of freedom, is $N - 1$.

8. The among-groups degrees of freedom, d.f.$_{among}$, is $(i - 1)$.

9. The error degrees of freedom, d.f.$_{error}$, is $(N - i)$.

10. The mean square among groups, $MS_{among} = SS_{among}/\text{d.f.}_{among}$.

11. The error mean square, $MS_{error} = SS_{error}/\text{d.f.}_{error}$.

12. The test statistic, $F = MS_{among}/MS_{error}$.

13. The degrees of freedom for F are d.f.$_{among}$, d.f.$_{error}$
    (note that F has *two* values for degrees of freedom, unlike other test statistics, which have only one).

14. Now look up the critical value of F in Table G of Appendix III for $f_1 = \text{d.f.}_{among}$, and $f_2 = \text{d.f.}_{error}$.

15. If your value for F is greater than the critical value, then the result is taken to be significant, and we reject the null hypothesis of equal mean values for all groups. Present the one-way ANOVA laid out in the standard manner (*see* Fig. (i)b in Box 3.5b [without, of course, the contrasts]).

16. Note that this is a *general* prediction and, therefore, we can only conclude that there are *some* differences among the means: we are *not* allowed to say which particular pair of mean values are significantly different because we did not make any specific prediction beforehand about the ordering of the mean values, nor did we set up planned specific contrasts between particular sets of groups. Many researchers would like to be able to make such *post hoc* tests, however, and a plethora of different methods are available and are often used. However, they should only be used as a rough guide to what the differences might be, preferably to set up a priori hypotheses for a new data set to test. Many people use the least significant difference, or LSD test. All this does is to take the MS_{error} from the ANOVA (which is an estimate of the variation within groups) and construct a 95 per cent confidence limit with it using:

$$LSD = t_{N-i} \times \sqrt{[(2/n) \times MS_{error}]}$$

where t_{N-i} is the 5 per cent threshold value of t for $N - i$ degrees of freedom. This looks a simple enough procedure, but in fact the LSD test is only accurate for a priori contrasts. It should, therefore, only be used as a rough guide in *post hoc* testing.

Box A3a	(ii) Mean values: how to do a specific parametric one-way ANOVA

Unfortunately there is effectively no parametric equivalent of the test for a particular rank order of mean ranks, as in Box A3b(ii). Although one does exist, called isotonic regression (*see* Gaines & Rice, 1990), it is very obscure and hardly ever used; it also involves computer-intensive randomisations of the data, and is well beyond the scope of this book. More common (but still unusual) is the use of a priori contrasts. If done in advance of obtaining the data, these allow $(i - 1)$ contrasts to be made. We explain this technique here.

1. Each contrast consists of one subset of the i groups (A) contrasted against another subset (B) (for example, a control (A) versus all treated groups (B)): each therefore involves effectively creating *two* groups out of the data, and testing them using a t-test. Such contrasts can themselves be either general ($A \neq B$) or specific (e.g. $A > B$).

2. Frame the predictions. Here you do this by formulating the contrasts that you want to make *in advance of collecting the data*. You are allowed $(i - 1)$ contrasts. Express each contrast as an inequality, and make the left-hand side greater than or equal to zero. For example, if you have three groups (A, B and C), and want to test whether group A is different from the other two groups (i.e. from the average of $B + C$), then this general hypothesis is that:

    $A \neq (B + C)/2$

    and hence

    $2A - B - C \neq 0$, i.e. $+2(A) - 1(B) - 1(C) \neq 0$

    This gives us a set of coefficients to apply to the mean values of each group; in this case they are +2, –1 and –1 for the mean values of groups A, B and C, respectively.

    Groups that you want to leave out of the contrast have a coefficient of zero. Thus another example might be that you are predicting that the control group A has a greater mean value than group C; this specific hypothesis is that:

    $A > C$

    and hence (making the left-hand side greater than zero)

    $+1(A) + 0(B) - 1(C) > 0$

    The coefficients are labelled as λ_i for the i groups.

3. You can check that you have a valid set of coefficients, because if so, then they sum to zero, i.e. $\Sigma\lambda_i = 0$.

    In addition, each contrast must be independent of all the others. This is the main reason why a maximum of $(i - 1)$ contrasts are allowed, since it is not possible for more than that to be independent. Statistically, if two contrasts are independent they are called orthogonal. If you have equal sample sizes per group (as every well-designed experiment should have!), you can easily check whether any two contrasts (*a* and *b*) are independent since the following must be true:

$$\sum \lambda_{ia} \times \lambda_{ib} = 0$$

In the case of unequal sample sizes of n_i per group, the equivalent formula is $\sum n_i \times \lambda_{ia} \times \lambda_{ib}$, and there are additional complications later on in the calculation. Strictly it is not essential for every contrast to be independent, but if they are not (as in this example), you need to adjust the threshold probability of significance: instead of 0.05, it becomes $1 - (0.95)^{1/r}$, where r = the number of non-independent contrasts you make.

4. Do a general parametric one-way ANOVA first (*see* Fig(i) in Box 3.5a), since technically you are breaking the among-group differences down into specific independent contrasts (*decomposing the sums of squares* in statistical jargon). You will need the error mean square, MS_{error}, to test each contrast.

5. For each contrast, obtain the coefficients (λ_i) and using the mean values (m_i) for each of the original groups, do the following calculation, which results in a sum of squares for the contrast. This contrast has one degree of freedom, and hence is also a mean square (*MS*):

$$L = \sum \lambda_i \times m_i$$

If you are testing a *specific* hypothesis about the contrast, check at this point that L is positive. If it is, this means that the data follow the predicted pattern (analogous to the positive value of a one-tailed *t*-test). If it is not positive, then you know already that your hypothesis will not be supported by the data, and will not be significant.

$$MS_L = L^2 n / \sum \lambda_i^2$$

where n is the sample size of each group.

If you have unequal sample sizes for your groups, follow the method of Sokal and Rohlf (1995, pp. 528–529), but you should probably use a computer package to do this calculation for you!

6. Now form the variance ratio

$$F = MS_L / MS_{error}$$

where MS_L is the mean square for the contrast, and MS_{error} is the error mean square from the one-way ANOVA. The test statistic is then

$$t = \sqrt{F}$$

7. The degrees of freedom of the *t*-test for the contrast are $a(n - 1)$, where a is the number of groups involved in the contrast (excluding the groups left out, those with coefficients of zero).

   (a) If your hypothesis is *general*, then look up the critical value of a two-tailed *t*-test in Table D of Appendix III.
   (b) If your hypothesis is *specific*, then look up the critical value of a one-tailed *t*-test in Table D of Appendix III.

8. Repeat steps 2–7 for each contrast.

9. Present the results of the one-way ANOVA, as well as the contrasts, laid out in the standard manner.

Box A3b (i) Mean ranks: how to do a general non-parametric one-way ANOVA

1. Formulate the prediction. In this case, the general prediction being made is to ask whether there are any differences at all among the mean ranks of the groups. Therefore the null hypothesis actually tested is that there are no differences among the group mean ranks.

2. The test considers i groups of data, each of which contains n_i data values. The total number of data values in all the groups together $= N = \sum n_i$.

3. Rank all the values across all the groups combined (as in the U-test), giving low rank scores to low values. Once again, tied values are given the average of the ranks they would have been ascribed had they been slightly different. Where there are lots of tied values relative to the sample size, you may need to apply a tie-correction factor, but in this case it is better to get the calculation done by a computer.

4. Sum the ranks in each group, giving R_i in each case.

5. The test statistic is H, where

$$H = \frac{12}{N(N+1)} \times (\sum R_i^2 / n_i) - 3(N+1)$$

6. The degrees of freedom are $(i - 1)$.

7. Look up the significance of the calculated H value as if it were a χ^2, in Table A of Appendix III (although H is not actually a χ^2, its value is distributed in the same way, so it is as if we were using χ^2). There is no standard layout for a general non-parametric ANOVA; just quote the test statistic, its degrees of freedom and the probability (*see* Fig. (i)b in Box 3.5b).

8. Note that this is a *general* prediction and, therefore, if significant we can only conclude that there are *some* differences among the mean ranks: we are *not* allowed to say which particular pair of mean values are significantly different because we did not make any specific prediction beforehand about the ordering of the mean values, nor did we set up planned specific contrasts between particular sets of groups. Many researchers would like to be able to make such *post hoc* tests, however, and some methods are available (but not often used). They should only be used as a rough guide to what the differences might be, preferably to set up a priori hypotheses for a new data set to test. Sokal & Rohlf (1995, p. 431) and Day & Quinn (1989) have some recommendations.

Box A3b (ii) Mean ranks: how to do a specific non-parametric one-way ANOVA

There are two ways of making specific predictions about the mean ranks of your groups. The one we favour here uses all the groups in a single a priori prediction of their rank order. The alternative is to use a priori contrasts (*see* point 9, below).

1. Formulate the specific prediction by specifying a particular rank order of the mean ranks of the groups, based on some a priori knowledge (theory, or previous published or gathered data), in advance of obtaining the data. The null hypothesis is that the rank order does not follow the prediction.

2. The test considers i groups of data, each of which contains n_i data values. There are N data values in total in all the groups ($= \Sigma n_i$).

3. Rank all the values across all the groups combined (as in the U-test), giving low rank scores to low values. Once again, tied values are given the average of the ranks they would have been ascribed had they been slightly different. Where there are lots of tied values relative to the sample size, you may need to apply a tie-correction factor, but in this case it is better to get the calculation done by a computer.

4. Sum the ranks in each group, giving R_i in each case.

5. Assign the predicted rank order to the groups, from the lowest (rank = 1) to the highest (rank = i). This rank order then provides the λ_i coefficient values. Using these λ_i values, calculate:

    the observed $L = \Sigma \lambda_i R_i$,

    the expected $E = (N + 1)(\Sigma n_i \lambda_i)/2$,

    the variance $V = (N + 1)(N \Sigma n_i \lambda_i^2 - (\Sigma n_i \lambda_i)^2)/12$.

6. Calculate the test statistic, z, as: $z = (L - E)/\sqrt{(V)}$. (Note that z is a standardised statistic which does not have any degrees of freedom.)

7. Look up the value of z in Table C of Appendix III, where you will see that the critical value for this specific (one-tailed) test is 1.64. If your value is greater than this, then the result is significant. If it is significant, we then reject the null hypothesis: there is evidence that the mean ranks fall into the predicted rank order.

8. What do you do if the result is *not* significant? If the mean ranks in fact fall in the opposite direction to your prediction, the value of z will be negative and quite possibly greater in absolute magnitude than 1.64. You *cannot* conclude *anything* about this, since your predicted rank order was not supported. You benefited from a gain in power over a general test, but the cost was that you could not conclude anything from a failure to reject the null hypothesis. You certainly cannot go and test an alternative rank order: this would now be *post hoc* since you have seen the actual pattern of the mean ranks. What you can do, following on from a non-significant specific test, is to ask the question: my predicted rank order was not supported, but is there evidence of *any* differences among groups in the data? In other words, you can go ahead and do a general test for any differences.

9. An alternative method is to use a priori contrasts, similar to the parametric case of Box A3a(ii). If done in advance of obtaining the data, these allow $(i - 1)$ independent contrasts to be made, each one consisting of a subset of the groups contrasted against another subset. Follow the method of Box A3a(ii), points 1–2, to obtain the coefficients for the contrast you want to make, and then create two new groups by adding together the data for all the original groups that have the same sign (+ or −) coefficient. These two artificial groups are then tested using either a Mann–Whitney U-test (general prediction only), or preferably a non-parametric one-way ANOVA (general or specific prediction).

    Note that the coefficients for these specific contrasts are *not* the same thing as the coefficients that specify the rank order for the test outlined above, points 1–8. You cannot use positive, negative and zero coefficients in the rank order test, imagining that you are doing a specific contrast.

10. There is no standard layout for a specific non-parametric one-way ANOVA; just quote the test statistic for each prediction, its degrees of freedom (if appropriate) and the probability. Note that a z-test does not have degrees of freedom, whereas specific contrasts have one degree of freedom (using either Mann–Whitney U-tests or non-parametric one-way ANOVA).

Box A4a (i) Mean values: how to do a general parametric two-way ANOVA

This is a little more complicated than a parametric one-way ANOVA, but in principle it is the same. Using the freshwater/marine fish example:

1. The two-way design is cast as ij cells (here 4) formed from i columns (here $i = 2$) and j rows (here $j = 2$) in a table. Each cell has a number of replicate measurements, n per cell.

    Formulate the predictions. In this case for example, (a) *rows*: marine fish differ in growth rate from freshwater fish; (b) *columns*: male fish differ in growth rate from female fish; (c) *interaction*: water type and sex of fish interact to determine growth rate. The appropriate null hypotheses are that there are no differences among the rows, or among the columns, and that there are no interactions of any kind.

2. Work out the mean values of all the replicates in each cell (B_{ij}), and the overall means of each column (C_i) and row (R_j) in the two-way design. The grand mean of all the data is M.

3. Calculate SS_{rows}, the sum of squares of the row mean values (not the raw data), following the method given in Box 3.1. Do the same for the column mean values to find SS_{cols}, and the cell means to find SS_{grps}.

4. Calculate the interaction sum of squares, $S_{int} = SS_{grps} - SS_{rows} - SS_{cols}$.

5. Calculate the sum of squares of each cell separately, and add them up to give SS_{error}.

6. The degrees of freedom for columns $d.f._{cols} = (i - 1)$, for rows $d.f._{rows} = (j - 1)$, for the interaction $d.f._{int} = (i - 1)(j - 1)$. The error degrees of freedom $d.f._{error} = ij(n - 1)$.

7. Calculate the mean squares: $MS_{rows} = SS_{rows}/d.f._{rows}$; $MS_{cols} = SS_{cols}/d.f._{cols}$; $MS_{int} = SS_{int}/d.f._{int}$; $MS_{error} = SS_{error}/d.f._{error}$.

8. Calculate the test statistic for each component:

    $F_{rows} = MS_{rows}/MS_{error}$ with degrees of freedom of $d.f._{rows}$, $d.f._{error}$

    $F_{cols} = MS_{cols}/MS_{error}$ with degrees of freedom of $d.f._{cols}$, $d.f._{error}$

    $F_{int} = M_{int}/MS_{error}$ with degrees of freedom of $d.f._{int}$, $d.f._{error}$

9. Look up these values of F with the appropriate degrees of freedom in Table G of Appendix III to see whether they exceed the relevant critical values. Present the two-way ANOVA laid out in the standard manner.

10. Note that if the interaction is significant, then the row and the column effects don't really mean much, because the difference among the row groups will then depend on which column group it is, and vice versa.

11. You should be aware that there are two kinds of grouping factors that create the 'ways' of a one- or two-way ANOVA. The distinction doesn't make a difference in a one-way parametric analysis, but it does in a two-way. First, there are *fixed* factors, where the groups are fixed by the experiment, and are not intended to represent a few of a great range of possibilities; the hypothesis being tested is about whether real differences exist among the actual groups in the experiment. Many experimentally created groupings are of this type, such as a treated and a control group (which are not representative of a range of different possible groups). A *random* factor, on the other hand, has groups that are merely a sample of all possible groups, and are intended to represent this range of possibilities; here one is not really interested in whether there are differences among these particular groups, but rather in how much of the variation is contained among as opposed to within the groupings. Rearing sets of animals in different cages would be a good example: cage differences are not really treatments, but merely random variation that is not interesting but nevertheless must be allowed for. Sokal & Rohlf (1995, Section 8.4) and Underwood (1997) have good discussions of this distinction.

Box A4a   (ii) Mean values: how to do a specific parametric two-way ANOVA

As in the one-way case (Box 3.5b), we cannot specify a particular rank order that we expect, but we can make a priori contrasts. If these contrasts are specified in advance of obtaining the data, they allow $(i-1)$ contrasts to be made among column groups, $(j-1)$ contrasts among row groups, and $(i-1)(j-1)$ contrasts involving the interaction of both row and column groups. As before, each contrast consists of a subset of the groups contrasted against another subset.

1. Formulate the predictions that you want to make in the form of contrasts, before collecting the data. You do this by expressing the prediction as an inequality, just as in Box 3.9b. From rearranging the inequality, you obtain the relevant set of coefficients, λ_i, for each contrast. Unlike Box 3.9b, however, you can either cast each contrast as a general or a specific contrast (using $\neq$ or $<$, respectively, in the inequality, just as in Box 3.5b, point 2).

2. Each contrast must be valid and independent of all the others (using the checks exactly as detailed in Box 3.5b, point 3).

3. Do a general parametric two-way ANOVA first (*see* Box 3.9a), since you are technically breaking the among-group differences down into specific independent contrasts ('decomposing the sums of squares' in statistical jargon). You will need the MS_{error} term from this analysis.

4. For each contrast, obtain the coefficients (λ_i) and the mean values (m_i) for each group, and follow exactly the method detailed for the one-way case in Box 3.5b, points 5 and 6, to obtain the test statistic, t, and its significance for your contrast.

5. Repeat step 4 for each contrast.

6. Present the results of the two-way ANOVA, as well as the contrasts, laid out in the standard manner.

Box A4b	(i) Mean ranks: how to do a general non-parametric two-way ANOVA

1. Rank the data values in all cells combined and add up the ranks in each cell to give a rank total R for that cell.

2. We can now use these rank totals to test our *general* predictions:

   (a) *Marine fish differ from freshwater fish.*
   Sum the rank totals for each *column* (*see table* on p. 105) separately giving an R_i value for marine and an R_i value for freshwater environments. Now calculate H as in the one-way analysis of variance (Box 3.4b), using the column R_i values and their appropriate n_i values (here there are eight values making up each R_i, hence $n_i = 8$), and check the resulting value in an appropriate table (e.g. Appendix III, Table A) as if it were χ^2 for $(i-1)$ degrees of freedom.

   (b) *Male fish differ from female fish.*
   Sum the rank totals for each *row* separately giving an R_i value for male and an R_i value for female fish and calculate H again as above. Again, $n_i = 8$ in our example.

   (c) *There is an interactive effect of water type and sex on the growth rate of fish.*
   This is an open-ended general prediction which combines both (a) and (b) above. It is asking whether there is any interaction between levels of grouping in determining growth rate. The calculation of H is exactly as above but the $\sum R_i^2/n_i$ term includes the rank totals for *all* the cells in the table instead of just the columns or the rows to give H_{tot}. Here n_i is again the number of values making up each R_i ($= 4$ in our example). H for the interaction, H_{int}, can then be calculated as: $H_{int} = H_{tot} - H_{water} - H_{sex}$, where H_{water} and H_{sex} refer to H values from (a) and (b) above. The degrees of freedom for H_{int} are $\text{d.f.}_{tot} - \text{d.f.}_{water} - \text{d.f.}_{sex}$, in this case, therefore, $3 - 1 - 1 = 1$ (d.f._{tot} is the total i (number of group means) minus 1).

Box A4b	(ii) Mean ranks: how to do a specific non-parametric two-way ANOVA

Testing our *specific* predictions is a little more complicated but still relatively straight-forward. We give three illustrations below:

(a) A prediction about the columns, e.g. *marine fish grow faster than freshwater fish* (or vice versa). In terms of the means in the table above this predicts that $(B + D) > (A + C)$. Another possibility is the opposite prediction, that freshwater fish grow faster than marine, i.e. $(A + C) > (B + D)$.

(b) A prediction about the rows, e.g. *male fish grow faster than female fish* (or vice versa). This predicts $(A + B) > (C + D)$ (or, conversely, for the prediction that females grow faster than males, $(C + D) > (A + B)$).

(c) A prediction about interaction, e.g. *the effect of water type on growth rate will be greater in male fish than in female fish.* This predicts that $(A - B) > (C - D)$. The converse (the effect will be greater in females) would, of course, predict $(C - D) > (A - B)$. This class of prediction is thus concerned with the *interaction* between water type and sex.

These are the predictions we can make about the relative sizes of the means; how do we arrive at the coefficients for testing? The procedure is as follows:

1. The first step is to rearrange the various predicted inequalities so that all the means are on the left, thus:

    (a) The prediction about the effect of water type becomes $-A + B - C + D > 0$ (for growth in marine > growth in fresh water) or $+A - B + C - D > 0$ (for growth in fresh water > growth in marine).

    We then substitute 1 with the appropriate sign for each letter so that we arrive at $-1, +1, -1, +1$ or $+1, -1, +1, -1$, respectively.

    (b) In the same way, the prediction about the effect of sex becomes $+A + B - C - D > 0$ (for males > females) or $-A - B + C + D > 0$ (for females > males) and the coefficients thus $+1, +1, -1, -1$ or $-1, -1, +1, +1$.

    (c) The interaction predictions must also be framed in this way. Thus the prediction $(A - B) > (C - D)$ becomes $+A - B - C + D > 0$ and the coefficients $+1, -1, -1, +1$. The prediction $(C - D) > (A - B)$ becomes $-A + B + C - D$ and the coefficients therefore $-1, +1, +1, -1$.

2. Then, testing these predictions

    (a) *Marine fish grow faster than freshwater fish*.
    Remember this means we are testing the prediction $-A + B - C + D > 0$. The coefficients λ_i thus become $-1, +1, -1, +1$ so that the rank totals for each cell are weighted as follows:

    $$L = \sum \lambda_i R_i$$
    $$= (-1)(R_{freshwater/male}) + (+1)(R_{marine/male})$$
    $$+ (-1)(R_{freshwater/female}) + (+1)(R_{marine/female})$$

    E and V can then be calculated as:

    $$E = (N + 1)(\sum n_i \lambda_i)/2$$
    $$V = (N + 1)[N\sum n_i \lambda_i^2 - (\sum n_i \lambda_i)^2]/12$$

    The test statistic z can then be calculated as:

    $$z = (L - E)/\sqrt{V}$$

    and checked against a table of z-values (*see* Appendix III, Table C).

    (b) *Male fish grow faster than female fish*.
    Now we are testing the prediction $+A + B - C - D > 0$, so λ_i becomes $+1, +1, -1, -1$. The calculation of L, E and V and then the test statistic z can proceed as above, but with the new λ_i weightings.

    (c) *The effect of water type is greater in males than in females*.
    This tests the interaction prediction $+A - B - C + D > 0$ using λ_i of $+1, -1, -1, +1$. Once again, follow the calculations above for L, E, V and z.

There is no standard layout for a non-parametric two-way analysis of variance; just quote the test statistic, its degrees of freedom and the probability.

Examples of tests for a trend

Box A5	How to calculate a correlation coefficient

Set out the two sets of data values to be correlated in pairs (remember, for each value in set 1 there must be a corresponding value in set 2). Thus, if we were looking for a correlation between height and weight in people, the data would be set out as below:

Person	Weight (kg)	Height (m)
1	63	1.8 (pair 1)
2	74	1.7 (pair 2)
3	60	1.9 (pair 3)
4	71	1.8 (pair 4)
.	.	. .
.	.	. .
.	.	. .
		etc.

It may be that several values of one measure are paired with the same value of the other, for example when measuring some behaviour in several individuals from the same social group and using these values in a correlation of time spent doing the behaviour and group size. In this case, the data might be as follows:

Observation	Time spent in behaviour (s)	Group size
1	15.3	3 (pair 1)
2	17.1	3 (pair 2)
3	18.0	5 (pair 3)
4	6.0	5 (pair 4)
5	31.1	5 (pair 5)
.	.	. .
.	.	. .
.	.	. .
		etc.

We can now calculate either a parametric (Pearson) or non-parametric (Spearman rank) correlation:

Box A5	(i) Pearson correlation coefficient (parametric)

This tests for a *linear* association between two (bivariate-) *normally* distributed variables.

1. Formulate the prediction, either as a general (any) or a specific (positive or negative) association. You can test for either a general (two-tailed) or specific (one-tailed) correlation coefficient by using different threshold values of the test statistic.

2. Calculate S_{xx}, S_{yy} and S_{xy} as shown in Box A1.

3. Calculate the test statistic, $r = S_{xy}/\sqrt{(S_{xx} \times S_{yy})}$.

4. Look up the threshold value for r using $(n-2)$ degrees of freedom in Table E of Appendix III, using either the one-tailed or two-tailed levels of significance, as appropriate to your hypothesis.

Box A5 (ii) Spearman rank correlation coefficient (non-parametric)

This tests for a *monotonic* (i.e. continuously increasing or decreasing) association between two variables, and works with ranking or constant interval measurements, making no assumptions about the normality of the data. As with the Pearson coefficient, we can test for either a general (two-tailed) or specific (one-tailed) correlation by using different threshold values of the test statistic.

1. Rank the values for the first measure only, then rank the values for the second measure only.

2. Subtract second-measure ranks from first-measure ranks (giving d_i) then square the resulting differences (d_i^2) and calculate the Spearman coefficient as:

    $$r_S = 1 - [(6\Sigma_i d_i^2)/(n^3 - n)]$$

    where n is the number of pairs of data values.

3. If n is between 4 and 20 (4 is a minimum requirement), consult Appendix III, Table F for the appropriate sample size to see whether the calculated r_S value is significant. Remember that general (two-tailed) or specific (one-tailed) tests have different threshold values for r_S. Thus you must know whether you are looking for any association at all (general), or just a positive, or just a negative one (specific).

4. If n is greater than 20, a different test statistic, t, is calculated from r_S as:

    $$t = r_S\sqrt{[(n - 2)(1 - r_S^2)]}$$

    t can then be checked against its own threshold values (Appendix III, Table D) for $n - 2$ degrees of freedom.

A significant calculated value for r_S or, with large samples, t, allows us to reject the null hypothesis of no trend in the relationship between our two measures.

Box A6 (i) How to do a parametric linear regression

1. Calculate S_{xx}, S_{yy} and S_{xy} as in Box A1.

2. Calculate the slope of the line as: $b = S_{xy}/S_{xx}$.

3. Calculate the intercept of the line on the y-axis as: $a = \bar{y} - b\bar{x}$, where $\bar{y}$ is the mean of the y-values and $\bar{x}$ is the mean of the x-values.

The line can now be fitted by calculating $y = a + bx$ for some sample x-values and drawing it on the scattergram.

To calculate the standard error of the slope:

4. Calculate the *variance of* y *for any given value of* x as:

    $$s_{y/x}^2 = [1/(n - 2)][S_{yy} - S_{xy}^2/S_{xx}].$$

5. The *standard error of the slope* is then: $\sqrt{(s_{y/x}^2/S_{xx})}$.

6. To find the test statistic F, calculate the following:

    Regression sum of squares (RSS) = $(S_{xy})^2/S_{xx}$
    Deviation sum of squares (DSS) = $S_{yy} - (S_{xy})^2/S_{xx}$
    Regression mean square (RMS) = RSS/1

    (1 is the value always taken by the *regression degrees of freedom*.)

    Deviation mean square (DMS) = DSS/$(n-2)$.

    ($n-2$ is the value taken by the *deviation degrees of freedom*.)

F is now calculated simply as F = RMS/DMS, and its value can be checked against critical values in F-tables (Appendix III, Table G) for 1 (f_1) and $n-2$ (f_2) degrees of freedom.

7. To find y for new values of x:
    Having established our regression equation, we might well want to predict y for other values of x that lie within the range we actually used in the analysis. Once we had then gone away and *measured* y for our new x-value we should want to see whether it departed significantly from its predicted value. Three steps are needed:

    (a) calculate the predicted y-value using the equation $y = a + bx$ as when fitting the regression line, but this time use the new x-value (x') in which you are interested;

    (b) calculate the standard error (s.e.) of the predicted y-value as follows:

    $$\text{s.e.} = \sqrt{[(s_{y/x}^2)(1 + 1/n + d^2/S_{xx})]}$$

    where $d = x' - \bar{x}$;

    (c) calculate the test statistic t as:

    $$t = \frac{\text{observed } y - \text{predicted } y}{\text{s.e.}}$$

    and look up the calculated value of t in t-tables (Appendix III, Table D) for $n-2$ degrees of freedom (where n is the number of pairs of data values in the regression). If t is significant it means the measured value of y departs significantly from the value predicted by the regression equation, and might lead to interesting questions as to why.

Box A6   (ii) How to do a non-parametric regression

A non-parametric approach to regression is little used (*see* Sokal & Rohlf, 1995: 539). There seem to be few advantages over using a simple rank correlation in cases where you do not want to predict a value of y from x, or to discover the actual equation of their relationship. However, some use is made of a non-parametric procedure (spline regression) fitted to a set of points to produce a smoothed description of highly non-linear irregular relationships (*see* Schluter, 1988; and Pentecost, 1999).

References

Day, R. W. & Quinn, G. P. (1989) Comparison of treatments after an analysis of variance in ecology. *Ecological Monographs* **59**(4): 433–463.

Gaines, S. D. & Rice, W. R. (1990) Analysis of biological data when there are ordered expectations. *American Naturalist* **135**(2): 310–317.

Pentecost, A. (1999) *Analysing environmental data*. Longman, Harlow.

Schluter, D. (1988) Estimating the form of natural selection on a quantitative trait. *Evolution* **42**: 849–861.

Sokal, R. R. & Rohlf, F. J. (1995) *Biometry*, 3rd edition. WH Freeman & Co., New York.

Underwood, A. J. (1997) *Experiments in ecology*. Cambridge University Press, Cambridge.

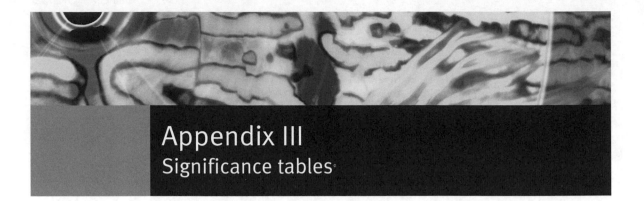

Table A Critical values of chi-squared at different levels of p. To be significant, calculated values must be *greater* than those in the table for the chosen level (0.05, 0.01, 0.001) of p and the appropriate number of degrees of freedom.

Degrees of freedom	Probability, p		
	0.05	0.01	0.001
1	3.841	6.635	10.83
2	5.991	9.210	13.82
3	7.815	11.34	16.27
4	9.488	13.28	18.47
5	11.07	15.09	20.51
6	12.59	16.81	22.46
7	14.07	18.48	24.32
8	15.51	20.09	26.13
9	16.92	21.67	27.88
10	18.31	23.21	29.59

Table B Critical values of Mann–Whitney U at $p = 0.05$. To be significant, values must be *smaller* than those in the table for appropriate sizes of n_1 and n_2.

n_1 \ n_2	3	4	5	6	7	8	9	10	15	20
2	–	–	0	0	0	0	0	0	1	2
3	0	0	1	2	2	2	2	3	5	8
4		1	2	3	4	4	4	5	10	13
5			4	5	6	7	7	8	14	20
6				7	8	10	10	11	19	27
7					11	12	12	14	24	34
8						15	15	17	29	41
9							17	20	34	48
10							20	23	39	55
15							34	39	64	90
20							48	55	90	127

Table C Probabilities associated with different values of z. The body of the table shows probabilities associated with different values of z. Values of z given to the first decimal place vertically and the second decimal place horizontally. z must therefore exceed 1.64 to be significant at $p < 0.05$.

z	.00	.01	.02	.03	.04	.05	.06	.07	.08	.09
1.5	.0668	.0655	.0643	.0630	.0618	.0606	.0594	.0582	.0571	.0559
1.6	.0548	.0537	.0526	.0516	.0505	.0495	.0485	.0475	.0465	.0455
1.7	.0446	.0436	.0427	.0418	.0409	.0401	.0392	.0384	.0375	.0367
1.8	.0359	.0351	.0344	.0336	.0329	.0322	.0314	.0307	.0301	.0294
1.9	.0287	.0281	.0274	.0268	.0262	.0256	.0250	.0244	.0239	.0233
2.0	.0228	.0222	.0217	.0212	.0207	.0202	.0197	.0192	.0188	.0183
2.1	.0179	.0174	.0170	.0166	.0162	.0158	.0154	.0150	.0146	.0143
2.2	.0139	.0136	.0132	.0129	.0125	.0122	.0119	.0116	.0113	.0110
2.3	.0107	.0104	.0102	.0099	.0096	.0094	.0091	.0089	.0087	.0084
2.4	.0082	.0080	.0078	.0075	.0073	.0071	.0069	.0068	.0066	.0064
2.5	.0062	.0060	.0059	.0057	.0055	.0054	.0052	.0051	.0049	.0048
2.6	.0047	.0045	.0044	.0043	.0041	.0040	.0039	.0038	.0037	.0036
2.7	.0035	.0034	.0033	.0032	.0031	.0030	.0029	.0028	.0027	.0026
2.8	.0026	.0025	.0024	.0023	.0023	.0022	.0021	.0021	.0020	.0019
2.9	.0019	.0018	.0018	.0017	.0016	.0016	.0015	.0015	.0014	.0014
3.0	.0013	.0013	.0013	.0012	.0012	.0011	.0011	.0011	.0010	.0010
3.1	.0010	.0009	.0009	.0009	.0008	.0008	.0008	.0008	.0007	.0007
3.2	.0007									
3.3	.0005									
3.4	.0003									
3.5	.00023									
3.6	.00016									
3.7	.00011									
3.8	.00007									
3.9	.00005									
4.0	.00003									

Table D Critical values of t at different levels of p. To be significant at the appropriate level of probability, values must be *greater* than those in the table for the appropriate degrees of freedom.

Degrees of freedom	Probability, p						
	0.05 0.10	0.025 0.05	0.01 0.02	0.005 0.01	0.001 0.002	0.0005 0.001	(one-tailed) (two-tailed)
1	6.314	12.71	31.82	63.66	318.3	636.6	
2	2.920	4.303	6.965	9.925	22.33	31.60	
3	2.353	3.182	4.541	5.841	10.21	12.92	
4	2.132	2.776	3.747	4.604	7.173	8.610	
5	2.015	2.571	3.365	4.032	5.893	6.869	
6	1.942	2.447	3.143	3.707	5.208	5.959	
7	1.895	2.365	2.998	3.499	4.785	5.408	
8	1.860	2.306	2.896	3.355	4.501	5.041	
9	1.833	2.262	2.821	3.250	4.297	4.781	
10	1.812	2.228	2.764	3.169	4.144	4.587	
11	1.796	2.201	2.718	3.106	4.025	4.437	
12	1.782	2.179	2.681	3.055	3.930	4.318	
13	1.771	2.160	2.650	3.012	3.852	4.221	
14	1.761	2.145	2.624	2.977	3.787	4.140	
15	1.753	2.131	2.602	2.947	3.733	4.073	
16	1.746	2.120	2.583	2.921	3.686	4.015	
17	1.740	2.110	2.567	2.898	3.646	3.965	
18	1.734	2.101	2.552	2.878	3.610	3.922	
19	1.729	2.093	2.539	2.861	3.579	3.883	
20	1.725	2.086	2.528	5.845	3.552	3.850	

Table E Critical values for the Pearson product–moment correlation coefficient r. Values must be *greater* than those in the table to be significant at the indicated level of probability.

Degrees of freedom	Probability, p					
	0.05 0.1	0.025 0.05	0.01 0.02	0.005 0.01	0.0005 0.001	(one-tailed) (two-tailed)
1	0.988	0.997	1.000	1.000	1.000	
2	0.900	0.950	0.980	0.990	0.999	
3	0.805	0.878	0.934	0.959	0.991	
4	0.729	0.811	0.882	0.917	0.974	
5	0.669	0.755	0.833	0.875	0.951	
6	0.622	0.707	0.789	0.834	0.925	
7	0.582	0.666	0.750	0.798	0.898	
8	0.549	0.632	0.716	0.765	0.872	
9	0.521	0.602	0.685	0.735	0.847	
10	0.497	0.576	0.658	0.708	0.823	
11	0.476	0.553	0.634	0.684	0.801	
12	0.458	0.532	0.612	0.661	0.780	
13	0.441	0.514	0.592	0.641	0.760	
14	0.426	0.497	0.574	0.623	0.742	
15	0.412	0.482	0.558	0.606	0.725	
16	0.400	0.468	0.543	0.590	0.708	
17	0.389	0.456	0.529	0.575	0.693	
18	0.378	0.444	0.516	0.561	0.679	
19	0.369	0.433	0.503	0.549	0.665	
20	0.360	0.423	0.492	0.537	0.652	
25	0.323	0.381	0.445	0.487	0.597	

Table F Critical values for the Spearman rank correlation coefficient r_S. Values must be *greater* than those in the table to be significant at the indicated level of probability.

	Probability, p				
	0.05	0.025	0.01	0.005	(one-tailed)
n	0.10	0.05	0.02	0.01	(two-tailed)
4	1.000				
5	.900	1.000	1.000		
6	.829	.886	.943	1.000	
7	.714	.786	.893	.929	
8	.643	.738	.833	.881	
9	.600	.700	.783	.833	
10	.564	.648	.745	.794	
11	.536	.618	.709	.755	
12	.503	.587	.671	.726	
13	.484	.560	.648	.703	
14	.464	.538	.622	.675	
15	.443	.521	.604	.654	
16	.429	.503	.582	.635	
17	.414	.485	.566	.615	
18	.401	.472	.550	.600	
19	.391	.460	.535	.584	
20	.380	.447	.520	.570	
21	.370	.435	.508	.556	
22	.361	.425	.496	.544	
23	.353	.415	.486	.532	
24	.344	.406	.476	.521	
25	.337	.398	.466	.511	

Table G Critical values of F at $p = 0.05$ (this page) and $p = 0.01$ (opposite). To be significant, values must be *greater* than those in the tables for the appropriate degrees of freedom (f_1 and f_2).

$p = 0.05$

f_2\\f_1	1	2	3	4	5	6	7	8	9	10	12	15	20	30	∞
1	161.4	199.5	215.7	224.6	230.2	234.0	236.8	238.9	240.5	241.9	243.9	245.9	248.0	250.1	254.3
2	18.51	19.00	19.16	19.25	19.30	19.33	19.35	19.37	19.38	19.40	19.41	19.43	19.45	19.46	19.50
3	10.13	9.55	9.28	9.12	9.01	8.94	8.89	8.85	8.81	8.79	8.74	8.70	8.66	8.62	8.53
4	7.71	6.94	6.59	6.39	6.26	6.16	6.09	6.04	6.00	5.96	5.91	5.86	5.80	5.75	5.63
5	6.61	5.79	5.41	5.19	5.05	4.95	4.88	4.82	4.77	4.74	4.68	4.62	4.56	4.50	4.36
6	5.99	5.14	4.76	4.53	4.39	4.28	4.21	4.15	4.10	4.06	4.00	3.94	3.87	3.81	3.67
7	5.99	4.74	4.35	4.12	3.97	3.87	3.79	3.73	3.68	3.64	3.57	3.51	3.44	3.38	3.23
8	5.32	4.46	4.07	3.84	3.69	3.58	3.50	3.44	3.39	3.35	3.28	3.22	3.15	3.08	2.93
9	5.12	4.26	3.86	3.63	3.48	3.37	3.29	3.23	3.18	3.14	3.07	3.01	2.94	2.86	2.71
10	4.96	4.10	3.71	3.48	3.33	3.22	3.14	3.07	3.02	2.98	2.91	2.85	2.77	2.70	2.54
11	4.84	3.98	3.59	3.36	3.20	3.09	3.01	2.95	2.90	2.85	2.79	2.72	2.65	2.57	2.40
12	4.75	3.89	3.49	3.26	3.11	3.00	2.91	2.85	2.80	2.75	2.69	2.62	2.54	2.47	2.30
13	4.67	3.81	3.41	3.18	3.03	2.92	2.83	2.77	2.71	2.67	2.60	2.53	2.46	2.38	2.21
14	4.60	3.74	3.34	3.11	2.96	2.85	2.76	2.70	2.65	2.60	2.53	2.46	2.39	2.31	2.13
15	4.54	3.68	3.29	3.06	2.90	2.79	2.71	2.64	2.59	2.54	2.48	2.40	2.33	2.25	2.07
16	4.49	3.63	3.24	3.01	2.85	2.74	2.66	2.59	2.54	2.49	2.42	2.35	2.28	2.19	2.01
17	4.45	3.59	3.20	2.96	2.81	2.70	2.61	2.55	2.49	2.45	2.38	2.31	2.23	2.15	1.96
18	4.41	3.55	3.16	2.93	2.77	2.66	2.58	2.51	2.46	2.41	2.34	2.27	2.19	2.11	1.92
19	4.38	3.52	3.13	2.90	2.74	2.63	2.54	2.48	2.42	2.38	2.31	2.23	2.16	2.07	1.88
20	4.35	3.49	3.10	2.87	2.71	2.60	2.51	2.45	2.39	2.35	2.28	2.20	2.12	2.04	1.84

Table G (Cont.)

$p = 0.01$

f_2 \ f_1	1	2	3	4	5	6	7	8	9	10	12	15	20	30	∞
1	4052	4999	5403	5625	5764	5859	5928	5982	6022	6056	6106	6157	6209	6261	6366
2	98.50	99.00	99.17	99.25	99.30	99.33	99.36	99.37	99.39	99.40	99.42	99.43	99.45	99.47	99.50
3	34.12	30.82	29.46	28.71	28.24	27.91	27.67	27.49	27.35	27.23	27.05	26.87	26.69	26.50	26.13
4	21.20	18.00	16.69	15.98	15.52	15.21	14.98	14.80	14.66	14.55	14.37	14.20	14.02	13.84	13.46
5	16.26	13.27	12.06	11.39	10.97	10.67	10.46	10.29	10.16	10.05	9.89	9.72	9.55	9.38	9.02
6	13.75	10.92	9.78	9.15	8.75	8.47	8.26	8.10	7.98	7.87	7.72	7.56	7.40	7.23	6.88
7	12.25	9.55	8.45	7.85	7.46	7.19	6.99	6.84	6.72	6.62	6.47	6.31	6.16	5.99	5.65
8	11.26	8.65	7.59	7.01	6.63	6.37	6.18	6.03	5.91	5.81	5.67	5.52	5.36	5.20	4.86
9	10.56	8.02	6.99	6.42	6.06	5.80	5.61	5.47	5.35	5.26	5.11	4.96	4.81	4.65	4.31
10	10.04	7.56	6.55	5.99	5.64	5.39	5.20	5.06	4.94	4.85	4.71	4.56	4.41	4.25	3.91
11	9.65	7.21	6.22	5.67	5.32	5.07	4.89	4.74	4.63	4.54	4.40	4.25	4.10	3.94	3.60
12	9.33	6.93	5.95	5.41	5.06	4.82	4.64	4.50	4.39	4.30	4.16	4.01	3.86	3.70	3.36
13	9.07	6.70	5.74	5.21	4.86	4.62	4.44	4.30	4.19	4.10	3.96	3.82	3.66	3.51	3.17
14	8.86	6.51	5.56	5.04	4.69	4.46	4.28	4.14	4.03	3.94	3.80	3.66	3.51	3.35	3.00
15	8.68	6.36	5.42	4.89	4.56	4.32	4.14	4.00	3.89	3.80	3.67	3.52	3.37	3.21	2.87
16	8.53	6.23	5.29	4.77	4.44	4.20	4.03	3.89	3.78	3.69	3.55	3.41	3.26	3.10	2.75
17	8.40	6.11	5.18	4.67	4.34	4.10	3.93	3.79	3.68	3.59	3.46	3.31	3.16	3.00	2.65
18	8.29	6.01	5.09	4.58	4.25	4.01	3.84	3.71	3.60	3.51	3.37	3.23	3.08	2.92	2.57
19	8.18	5.93	5.01	4.50	4.17	3.94	3.77	3.63	3.52	3.43	3.30	3.15	3.00	2.84	2.49
20	8.10	5.85	4.94	4.43	4.10	3.87	3.70	3.56	3.46	3.37	3.23	3.09	2.94	2.78	2.42

Appendix IV
The common codes for the important graphical parameters of R

col =	lty =	pch =
0 blank	blank	open square
1 black	solid	open circle
2 red	dashed	open up triangle
3 green	dotted	plus
4 blue	dotdash	X
5 cyan	longdash	open diamond
6 magenta	twodash	open down triangle
7 yellow		X in box
8 grey		star
9		cross in diamond
10		cross in circle
11		6-point star
12		cross in box
13		X in circle
14		triangle in box
15		filled square
16		filled circle
17		filled up triangle
18		filled diamond

The general topic and page number relevant to an answer is given in brackets in each answer.

1. No, this is not an appropriate analysis because it involves the multiple use of a two-group test (*Mann–Whitney U-test Box 3.3c* pp. 76–77). A significant result becomes more likely by chance the greater the number of two-group comparisons that are made. An appropriate analysis would be a non-parametric one-way analysis of variance (*Box 3.5d*, pp. 92–95), which allows a test of the specific prediction that body size will be greatest in Lake 1, intermediate in Lake 2 and least in Lake 3.

2. Regression analysis (*Box 3.11* pp. 127–129) is appropriate when testing for cause and effect in the relationship between *x*- and *y*-values, and the *x*-values are established in the experiment. The data are measured on some kind of constant interval scale that allows a precise, quantitative relationship to be calculated. Because it depends on establishing a quantitative relationship, predicting new values of one variable from new values of the other also demands regression analysis (*Linear regression* p. 124). In other cases, correlation analysis is necessary. Non-parametric correlation analysis (*Box 3.10*, pp. 122–123) is appropriate for both these kinds of data, and others where *x*-values are merely measured, but yields only the sign and magnitude of the relationship and makes no assumptions about the cause-and-effect relationship between *x*- and *y*-values (*Cause and effect* p. 125). (See also: *What sort of trend?* p. 201)

3. The investigator has introduced a new analysis into the Discussion (the correlation between nutrient flow rate and aphid production). The analysis should, of course, be in the Results section (*Results* pp. 171–172; see especially last sentence of this section. See also *Discussion* p. 172, third sentence).

4. The information tells you that the investigator carried out a non-parametric one-way analysis of variance and that they tested a general prediction, thus using the test statistic H instead of z. The three degrees of freedom tells you that the analysis compared four groups and the p-value that $H = 14.1$ at d.f. = 3 is significant at the 1 per cent level. (*Box 3.5c*, pp. 91–92)

5. The researcher can conclude that there is a significant positive association between daily food intake and growth rate, but can't necessarily infer that increased growth rate is caused by greater food intake; it could be that faster-growing pigs simply eat more food as a result. (*Cause and effect* p. 125)

6. No, a chi-squared test is not appropriate here because the data are constant interval measurements and not counts. (*Types of measurement and types of test* pp. 57–59)

7. (a) A test statistic is calculated by a significance test, and its value has a known probability of occurring by chance for any given sample size or number of degrees of freedom. (*Test statistic* p. 68)

   (b) A ceiling effect occurs where observational or experimental procedures are too undemanding to allow a prediction to be tested; all samples approach the maximum value. (*Floor and ceiling effects* p. 138)

   (c) Statistical significance refers to cases where the probability that a difference or trend as extreme as the one observed could have occurred by chance, if the null hypothesis of no difference or trend is true, is equal to or less than an accepted threshold probability (usually 5 per cent, but sometimes 1 or 10 per cent). (*What is statistical significance?* pp. 54–56)

8. No, the observer cannot conclude that male thargs prefer larger females just from this. It could be, for instance, that larger females are simply more mature and thus more likely to conceive. Alternatively, depending on how and when size was measured, pregnant females may be larger precisely because they are pregnant! (*Cause and effect* p. 125)

9. Significance tests provide a generally accepted, arbitrary yardstick for deciding whether a difference or trend is interesting. The yardstick is the probability that the observed difference or trend could have occurred by chance when there wasn't really such a difference or trend in the population. Random variation in sampling will mean that differences or trends will crop up from time to time just by chance. (*The need for a yardstick in confirmatory analysis: statistical significance* pp. 52–54)

10. A reasonable prediction would be that the rate of reaction would be lowest in Treatment A, because no enzyme had been added, and highest in the warmed enzyme/substrate mixture of Treatment D. By the same rationale, Treatment C should have a lower rate than Treatment B because it is cooler. The predicted order is thus A < C < B < D. A suitable significance test would be a specific form of a non-parametric one-way analysis of variance using z as the test statistic. (*Testing rank order* pp. 118–119)

11. The consultant could try a two-way analysis of variance (*Box 3.9a* pp. 106–109 or *Box 3.9c* pp. 116–117). The two levels of grouping (*Levels of grouping* p. 105) would be 'housing condition' and 'breed', with three groups at the first level and four at the second. A dozen or so samples for each combination of housing and breed would be useful, though the same number of samples should be used in each case since general rather than specific predictions are being tested. One thing the consultant should be careful to do is distribute pigs from different families arbitrarily across housing conditions so that any effect of housing on growth rate is not confounded with family-specific growth rates (related pigs might grow at a similar rate because they inherited similar growth characteristics) (*Confounding effects* pp. 134–138). The analysis would indicate any independent effect of housing and breed and any interaction between the two in influencing growth rate.

12. Although the figures look very similar, they are deceptive because their *y*-axes are scaled differently (*see point 2* p. 159). The drop in numbers in felled deciduous forest represents 49 per cent of the number in unfelled forest. In coniferous forest it represents only 38 per cent. In proportional terms, therefore, the impact of felling seems to be greater in deciduous forest. However, the fact that the analysis is based on only single counts means it should act only as an exploratory analysis leading to a properly replicated confirmatory analysis. (*Visual exploratory analysis* pp. 20–23)

13. A negative value of *r* indicates a negative correlation, i.e. the value of *y* decreases as that of *x* increases. The sign of the coefficient is ignored when checking against threshold values of significance. (*Correlation analysis* pp. 121–124)

14. Some predictions can be derived as follows:

(a) *Difference predictions*

Observation The distribution of individuals across species varied between different sites.

Hypothesis *Differences in the degree of dominance of species within a community vary with the ability of species to compete with others for limited resources.*

Prediction *Dominant species will be those whose individuals win in contests with individuals of other species over the resource they occupy.*

Observation Individuals of some species are smaller in some streams than in others, and some streams are more polluted.

Hypothesis *Pollution results in reduced body size among some freshwater species.*

Prediction *The body size of any given species will be smallest in the most polluted stream and largest in the least polluted stream.*

(b) *A trend prediction*

Observation Fewer organisms were seen attached to the substrate in fast-flowing parts of the streams.

Hypothesis *Flow rate influences the ability of organisms to settle on the substrate.*

Prediction *If clean substrate is provided in areas of different flow rate, then fewer of the organisms drifting by will settle the faster the flow.*

(*Turning exploratory analyses into hypotheses and predictions* p. 42–46)

15. Differences. (*Differences and trends* pp. 47–50; *Difference or trend?* pp. 201)

16. Yes, a 1×4 chi-squared analysis (*Box 3.7* p. 103) could be carried out, but care would have to be taken in calculating expected values because of the different surface areas of the body sites and their possibly different degrees of vulnerability. (*Null hypotheses* pp. 46–47)

17. Since the experimenter had collected ten counts for each combination of 'parent' and 'sibling' treatment, a two-way analysis of variance would provide most information. It would allow either general or specific predictions about the effects of 'parent' and 'sibling' treatments and the interaction between them to be tested. The ten values in each case could be totalled and used in a 2×2 chi-squared analysis, but this would test only for an over-all combined effect of the treatments; much useful information would thus be lost in comparison with the analysis of variance. (*Box 3.9a* pp. 106–109 or *Box 3.9c* pp. 116–117 versus *Box 3.8* p. 104)

18. The threshold probability of 0.05 is an arbitrarily agreed compromise between the risk of accepting a null hypothesis in error (as would happen by setting the threshold *p*-value too high) and the risk of rejecting it in error (by setting the threshold too low). For many situations in biology, a threshold of $p = 0.01$ would result in an unnecessary risk of accepting the null hypothesis when it was not true. (*What is statistical significance?* pp. 54–56)

19. Clearly this is not a sensible procedure because it confounds the size of prey with the amount each barracuda has already eaten. Barracuda may give up at a certain size of fish simply because they are satiated, not because the fish is too big. Also, size is confounded with species, some of which may be distasteful or be unpalatable in other ways. Again, therefore, barracuda may reject a fish for reasons other than size. (*Confounding effects* pp. 134–138)

20. There are at least three logical flaws in this line of reasoning. First, if males are generally larger than females then their brains will be proportionally larger too; any comparison of brain sizes should thus be on a relative scale. Second, should the brains of males turn out to be relatively larger, there is no a priori reason to suppose this will affect learning or any other ability. Third, even if differences in brain size do produce differences in learning, training may overcome any such differences. (*Refining hypotheses and predictions* pp. 145–146)

21. The analysis requires a test for a difference (*Difference or trend?* p. 201). There is one level of grouping (different woods), so a one-way analysis of variance (*Box 3.5a* pp. 84–86 or *Box 3.5c* pp. 91–92) of the effect of wood on adrenal gland size would be appropriate. The potentially confounding factor (*Confounding effects* pp. 134–138) of age may be a problem, but as long as the distributions of adrenal gland size and age are normal (*Parametric tests* pp. 58–59), the ecologist could do a parametric analysis of variance and include age as a covariate (*Box 3.6* pp. 100–102). The result would then test for an effect of wood, controlling for any confounding effect of age.

22. An order effect may occur when there is a systematic confounding of experimental treatment and time. Thus, testing for a difference in the performance of chicks in three different learning environments would suffer from an order effect if birds always encountered environment 1 first, then environment 2 and then environment 3. Any difference between environments may simply be due to being earlier or later in the sequence experienced by the birds (or earlier or later in the day/week/season if individual birds experienced only one of the environments). (*Confounding effects* pp. 134–138, particularly last three sentences on p. 138)

23. On the face of it, things don't look good for the parasitologist's regression analysis. The parasite-burden data clearly aren't normal, and so violate an important assumption about the y-axis variable in a linear regression. However, all is not lost. It may be possible to normalise the data by performing an appropriate transformation. Indeed, a $\log_{10}$ transformation normalises the data quite nicely, as the figure overleaf shows (a Kolmogorov–Smirnov one-sample test yielded the outcome $z = 1.113$, $n = 120$, $p = 0.157$, showing the parasite data did not depart significantly from a normal distribution). The parasitologist can therefore carry on with the planned analysis using the transformed data. (*Departures from normality* p. 60, *Transformations* pp. 60–61, *Box 3.1b* pp. 63–65)

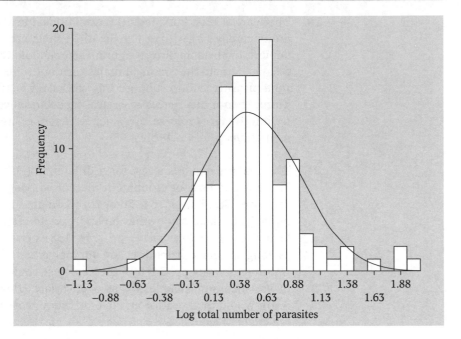

24. Pseudoreplication occurs where there is non-independence between measures purporting to be independent replicates. An extreme case would be taking repeated measurements of a particular character from the same individual, but more subtle pseudoreplication can arise, for example, when animals from the same cage or litter are treated as independent samples. Animals sharing a cage can have a profound influence on each other behaviourally, physiologically and even morphologically, and those from the same litter obviously share a genetic and experiential background. The number of cages or litters, rather than the number of individuals, thus determines the sample size in any statistical analysis. (*Non-independence* pp. 138–139)

25. A bit sneaky this one. Unlike most significance tests of difference, you are looking for a *non-significant* outcome when you compare the distribution of data with a normal distribution. If the comparison is not significant, it means your data do not depart from normality more than you would expect by chance. In the example the probability associated with the test outcome was 0.0341, which is less than 0.05. The data thus differ significantly from normal, so the botanist will have to use transformed data if they want to do a parametric significance test, or use a non-parametric test instead. (*Box 3.1b* pp. 63–65, particularly p. 64)

26. The behavioural ecologist is testing for a difference between three experimental treatments (since the treatments involve the number of males, it could also be argued to be testing for a trend, but it is not clear a priori that any response by focal males would relate linearly to the number of

competitors, so the analysis is probably best treated as one of difference). So an analysis of variance (ANOVA) of some kind would be the obvious choice. A crucial feature of the experimental design, however, is that each focal male was tested in each of the treatments, so we have a *repeated measures* design. Assuming data for spermatophore weight satisfy the test for normality, we can use a parametric repeated-measures ANOVA (*Box 3.5e* pp. 95–97). Since males may be tempted to invest more when confronted with larger females (because they are likely to have more eggs to fertilise), female thorax widths could be included as a covariate (*Box 3.6* pp. 100–102) in the analysis to control for this. Since the design is a repeated-measures one, we do not have to control for male size – the treatment effects are being measured within each male. If spermatophore weight is not normally distributed, then a non-parametric repeated-measures Friedman ANOVA (*Box 3.5f* pp. 98–100) could be used, but it isn't possible to include covariates in this case.

27. An obvious problem facing the psychologist is that the questionnaire data relating to attitudes will contain a lot of responses reflecting different aspects of sexual behaviour, many of which may be intercorrelated. So carrying out an analysis of each response individually would run into serious problems of non-independence and inflated estimates of significance (*Non-independence* pp. 138–139). One approach to overcome this would be to subject the 40 responses to a principal components analysis (*Box 3.13* pp. 132–133) to distil the data down to a set of mutually uncorrelated composite variables that each reflect different aspects of attitudes towards sex. The 40 separate responses may well boil down to two or three composite variables that can then be used as dependent variables in an analysis of variance by, for example, sex and ethnic group, with perhaps financial status (e.g. salary) and age as covariates (*Box 3.6* pp. 100–102).

28. As long as clearance rate is normally distributed, the idea of using regression analysis is reasonable, but what is problematic is the use of four separate analyses. The independent variables are very likely to be intercorrelated (e.g. maternal weight may well depend on local food availability), so analysing them as if they are all independent of each other risks a Type 1 error (*Box 3.14* pp. 135–137). A better approach would be to conduct a multiple regression where the independent variables are all included in a single analysis and their mutual intercorrelations taken into account in assigning significance. (*Box 3.12* pp. 129–131)

29. Since many behavioural variables were recorded that are (probably) intercorrelated, the best approach would be to perform a principal components analysis (*Box 3.13* pp. 132–133) first to reduce the many variables to their main features. It is likely that this will produce two or three components (each a composite of the original variables) that can be interpreted in terms of combinations of the bees' responses. Each of these components can be used as a dependent variable in a one-way analysis of variance of flower

treatment (with three levels: unmanipulated control, previously visited by the subject bee, previously visited by a different bee). (*Box 3.5a* pp. 84–86 or *Box 3.5c* pp. 91–92)

30 The treatment has four levels (three strengths of electromagnetic field and a control), and apart from the control, no information is available about the relative magnitudes of the experimental fields. Each response measure is independent of the others because each neurone preparation was used only once; it is thus clearly not a repeated-measures design. A simple one-way analysis of variance would thus be appropriate. The only issue is whether a general or a specific prediction should be tested (e.g. *Box 3.5a* pp. 84–86 or *Box 3.5b* pp. 86–90). Unless there is a good a priori reason for expecting a particular directional effect of field strength, the only possible specific prediction is the contrast of control versus electromagnetic levels (i.e. coefficients of +3 [control] and –1 for each treatment): this is a sensible choice of specific contrast. (*Box 3.5b* pp. 86–90)

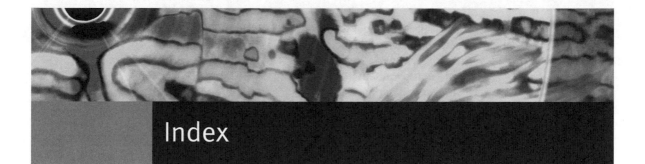

Index

Quick Test Finder: Tests for Difference

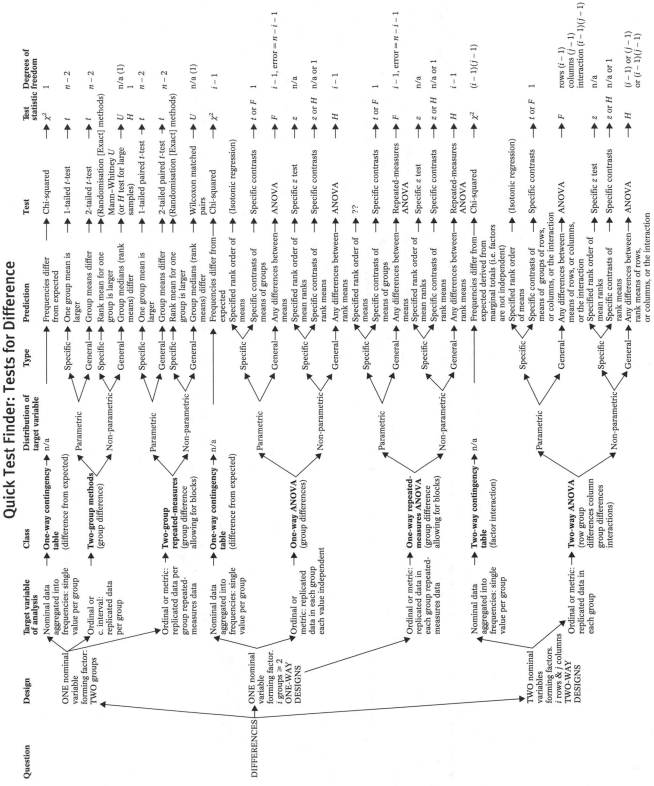

Quick Test Finder: Tests for Trends

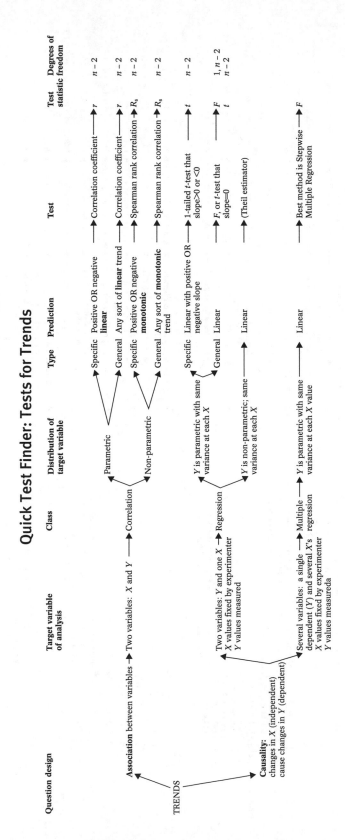

Question design	Target variable of analysis	Class	Distribution of target variable	Type	Prediction	Test	Test statistic	Degrees of freedom
Association between variables	Two variables: X and Y → Correlation		Parametric	Specific	Positive OR negative **linear**	Correlation coefficient →	r	$n - 2$
				General	Any sort of **linear** trend	Correlation coefficient →	r	$n - 2$
			Non-parametric	Specific	Positive OR negative **monotonic**	Spearman rank correlation →	R_s	$n - 2$
				General	Any sort of **monotonic** trend	Spearman rank correlation →	R_s	$n - 2$
Causality: changes in X (independent) cause changes in Y (dependent)	Two variables: Y and one X. X values fixed by experimenter. Y values measured → Regression		Y is parametric with same variance at each X	Specific	Linear with positive OR negative slope	1-tailed t-test that slope>0 or <0 →	t	$n - 2$
				General	Linear	F, or t-test that slope=0 →	F; t	$1, n - 2$; $n - 2$
			Y is non-parametric; same variance at each X		Linear	(Theil estimator) →		
	Several variables: a single dependent (Y) and several X's. X values fixed by experimenter. Y values measureda → Multiple regression		Y is parametric with same variance at each X value		Linear	Best method is Stepwise Multiple Regression →	F	

TRENDS